# Lichens

Wi 1liam Purvis

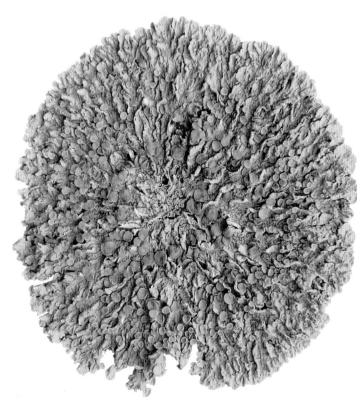

SMITHSONIAN INSTITUTION PRESS, WASHINGTON, D.C.
IN ASSOCIATION WITH THE NATURAL HISTORY MUSEUM, LONDON

Dedicated to Peter James who has done so
much to stimulate lichenology worldwide

Published in the United States of America
by the Smithsonian Institution Press
in association with
The Natural History Museum
London
Cromwell Road
London SW7 5BD
United Kingdom

Library of Congress Cataloging-in-Publication Data
Purvis, William, (Ole William).
        Lichens / William Purvis.
                p. cm.
        Includes bibliographical references (p.  ).
        ISBN 1-56098-879-7 (alk. paper)
        1. Lichens. I. Title.

        QK583.P87  2000
        579.7–dc21                              00-029146

Manufactured in Singapore, not at government expense
07 06 05 04 03 02 01 00  5 4 3 2 1

Edited by Jacqui Morris
Designed by David Robinson
Reproduction and printing by Craft Print, Singapore

DISTRIBUTION

North America, South America,
Central America, and the Caribbean
Smithsonian Institution Press
470 L'Enfant Plaza
Washington, D.C. 20560-0950
USA

Australia and New Zealand
CSIRO Publishing
PO Box 1139
Collingwood, Victoria 3066
Australia

UK and rest of the world
Plymbridge Distributors Ltd.
Plymbridge House, Estover Road
Plymouth, Devon PL6 7PY
UK

# Contents

Preface                                          4

What is a lichen?                                5

How lichens grow, multiply and disperse         19

Lichen biodiversity                             33

Evolution, classification and naming            46

Ecological role                                 49

Lichens in forests                              56

Lichens in extreme environments                 62

Biomonitoring                                   76

Prospecting and dating                          88

Economic uses                                   92

Practical projects                              99

Glossary                                       108

Index                                          109

Further information                            111

Acknowledgements                               112

# Preface

This book explores the fascinating world of lichens. Lichens occur almost everywhere and the lichen symbiosis is one of the most successful examples of two or more organisms living together. How lichens are formed is one of the greatest puzzles in biology. Just as salt is dissimilar to sodium and chlorine, so is a lichen dissimilar to the organisms that create it. Lichen morphology, physiology and biochemistry are very different to that of the isolated fungus and alga in culture. Lichens occur in some of the most extreme environments on Earth – arctic tundra, hot deserts, rain forests, rocky coasts and toxic spoil heaps. Many of these environments are also among the most sensitive to destruction and pollution. Our rich and varied lichen heritage is incredibly useful in allowing us to assess what we are doing to our environment.

Lichenology is at an exciting stage of development, attracting great interest from biologists, chemists, geologists, environmental scientists and consultants, and industry. Lichens are used successfully as bioindicators of air quality by school children and students throughout the world. The rewards of understanding lichen biodiversity are great and in the future may lead to important technological developments, with the discovery of new medicines and the clean-up of contaminated sites, as well as monitoring the health of our environment.

# The author

William Purvis graduated in botany from the University of Sheffield in 1981 and was awarded a Natural History Museum (NHM) Centenary studentship funded by British Petroleum at the Royal School of Mines, Imperial College and Botany Department, The Natural History Museum, London. He took up his post as a lichenologist at The Natural History Museum in 1988. He is principal author of the *Lichen Flora of Great Britain and Ireland* and has carried out pioneering research into why different lichen species accumulate metals, and the implications for biomonitoring and classification. He is actively involved in using lichens as bioindicators of environmental health and is carrying out taxonomic research on critical groups.

# What is a lichen?

Many people think that lichens are simple organisms rather like mosses. In fact, they are mini-ecosystems, consisting of at least two organisms: a fungus (mycobiont) and a photosynthetic partner (photobiont). The photobiont, which contains chlorophyll, may be either a green alga or belong to an entirely different kingdom – a cyanobacterium (a bacterium that contains a blue-green photosynthetic pigment). Even where only two partners are present, the association or symbiosis is not usually a simple mixture. Instead, it results in a distinct lichen body

LEFT **Lichen-covered fence post, with leafy orange** *Xanthoria* **sp., shrubby** *Usnea* **sp. (green thalli) and** *Hypogymnia* **sp. (pale-grey, left) and a crust-like** *Thelomma* **sp. (black-and-white, top). Coastal hills of Northern California.**

ABOVE **Boulders with lichens (***Stereocaulon*** spp. and reindeer lichens,** *Cladonia* **spp.), mosses (***Rhacomitrium***,** *Andreaea* **and** *Polytrichum* **spp.) and shrub. Parc Grandes Jardins, Laurentides, Quebec.**

(thallus) being formed. It looks very different from the individual partners when they are separated and grown in laboratory cultures. The names we call lichens refer to the fungal partner because every lichen has a unique fungus. The algae and cyanobacteria have their own names and most of them occur in many different lichens.

Fungi, and algae and/or cyanobacteria associate in many different ways in lichens. At one end of the scale fungal filaments and algae coexist, as in the tropical lichen genus *Coenogonium* (see p. 45). At the other end of the scale are the majority of lichens, which are highly organized and, without detailed examination, appear to be a single entity. The International Association of Lichenology in 1982 defined a lichen as an "association of a fungus and a photosynthetic symbiont resulting in a stable thallus of specific structure". However, where several partners occur in the same thallus, the precise role played by each individual partner is by no means clear.

An intriguing question is whether the mycobiont and the photobiont benefit from the association. Many biologists consider lichens to be one of the finest examples of symbiosis (two or more organisms living together) because lichens are widespread and the algal cells appear to be healthy inside the lichens. Others believe that the lichen relationship is one of controlled parasitism, that the photobiont cells are victims rather than partners of the mycobiont. Simon Schwendener, a German, was the first to discover the dual nature of lichens. In 1869, he wrote:

RIGHT **Section through** *Heppia adglutinata* **showing algal cells (from Schwendener 1869).**

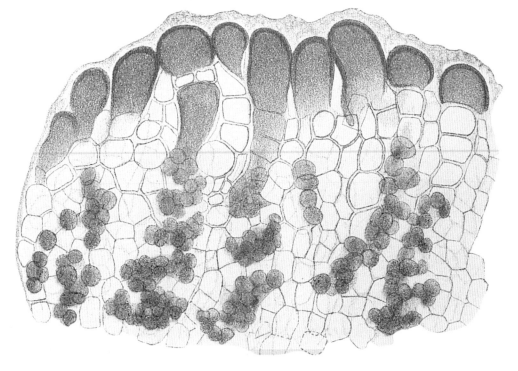

As the result of my researches, the lichens are not simple plants, not ordinary individuals in the ordinary sense of the word; they are, rather, colonies, which consist of hundreds of thousands of individuals, of which, however, one alone plays the master, while the rest, forever imprisoned, prepare the nutriment for themselves and their master. This fungus is a fungus of the class Ascomycetes, a parasite which is accustomed to live upon others' work. Its slaves are green algae, which it has sought out, or indeed caught hold of, and compelled into its service. It surrounds them, as a spider its prey, with a fibrous net of narrow meshes, which is gradually converted into an impenetrable covering, but while the spider sucks its prey and leaves it dead, the fungus incites the algae found in its net to more rapid activity, even to more vigorous increase.

Schwendener's views were quite revolutionary for the time, and his dual hypothesis, as it was called, was not accepted universally. Indeed, it was rejected with scorn by some of the most prominent lichenologists of the day. One of these, the Scottish Reverend James Crombie, described it as "a romance of lichenology, or the unnatural union between a captive algal damsel and tyrant fungal master". It is easy to see how the lichen fungus benefits from the association – the photobionts photosynthesize and provide it with carbohydrate. This is in the form of sugar alcohols in the case of lichens containing green algae and glucose in lichens containing cyanobacteria. The mycobiont may also obtain nitrogen compounds when the photobiont is a cyanobacterium because cyanobacteria can capture and fix nitrogen from the atmosphere.

We do not have any evidence that the fungus passes any nutrients to the photobiont, but it does modify the algae and cyanobacteria that it contains. The photobionts lose their cell walls and do not reproduce sexually. However, both partners share one benefit. Together they are able to colonize many habitats that would otherwise be out of their reach. The fungus can live in places lacking the organic matter that they would normally need as a source of nutrients. Algae and cyanobacteria, which usually live in aquatic or moist habitats, can live in drier places. They can also be affected adversely by high light intensity and, given the protection of the fungus, they can expand into environments where light intensity is high.

Symbiosis is a very successful way of living, but it is not unique to lichens. Other examples are:

• Mycorrhizas – associations of fungi with higher plants, in which the fungus lives on the plant roots.

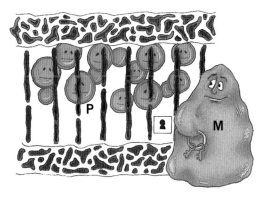

LEFT **Is the lichen thallus a controlled form of parasitism? M, mycobiont; P, photobiont.**

• Actinorrhizas – where bacteria are associated with the roots of leguminous plants.

• Freshwater and marine invertebrates – *Hydra* and some other coelenterates include algae in their bodies, as do some flatworms.

# Lichen fungi

Fungi, unlike green plants, contain no chlorophyll and cannot produce their own carbohydrates. They have developed various ways of obtaining organic carbon compounds from both living and dead sources (heterotrophy). Fungi form many different symbiotic relationships with vascular plants (mycorrhizas). For instance, fungi may be associated with forest trees, where they assist in the nutrition of their host. The process of lichenization, involving obtaining carbon from photobiont cells, is a common mode of fungal nutrition; around 20% of all fungi are lichenized. Many lichen fungi produce special microscopic branches (haustoria), which penetrate the photobiont. The lichen-forming fungi are a very mixed group. Some belong to taxonomic groups in which all the members are lichenized, while other groups may contain non-lichenized relatives. By far the largest proportion of lichen-forming fungi (over 40%) belong to the ascomycetes, which produce spores in a sac-shaped cell (ascus).

In nature lichen partners exist only in symbiosis, apart from as a reproductive spore. In laboratory conditions, however, it is possible to take the lichen apart and grow the two partners separately. Scientists in Europe,

LEFT **Orange-peel fungus, *Aleuria aurantia* – a non-lichenized ascomycete fungus. Epping Forest, London, UK.**

Japan and North America have made significant progress in culturing the fungus from spores or extracts of fungal cells. In pure culture, lichen mycobionts form slow-growing amorphous colonies, which are quite different in appearance from the whole (lichenized) thallus. It is also possible to culture some lichens and to synthesize new ones by inoculating spores with algae, but they are difficult to maintain for long periods. Success depends partly on avoiding contaminant organisms killing the lichen and also on creating the right conditions, which do not favour one or other of the symbionts. Solid (agar) culture media are better for growing lichens than liquid media.

## Lichen algae

Only a few types of algae occur within lichens. Algae can be the dominant partner, e.g. in the 'jelly lichens', *Collema* spp. and *Leptogium* spp., but this is rare. The algae found most frequently are the unicellular green algae of the genus *Trebouxia*. Other common species belong to the orange-pigmented genus *Trentepohlia* and the

LEFT *Xanthoria parietina*: the orange, disc-like fruiting bodies of this lichen are superficially similar to those of the fungus *Aleuria aurantia*.

ABOVE The fungus of *Xanthoria parietina* in culture. It is amorphous and looks quite unlike the intact lichen. Lichen fungi are not known to produce fruiting bodies in culture.

9

cyanobacterium (formerly called blue-green alga) genus *Nostoc*. Lichen photobionts look rather different when they are inside rather than outside lichens. Sexual reproduction and the formation of filaments is suppressed and we usually need to isolate and culture lichen algae to identify them to species level. This has been achieved successfully for less than 2% of all lichen photobionts. Around 100 green algae (occurring in about 23 genera) and many fewer cyanobacterial species in about 15 genera may be involved as symbionts. Most lichen fungi are highly selective, forming associations only with compatible photobiont species. Some notable exceptions include:

• The same lichen in different geographical locations may contain different photobionts.
• Some lichen species contain different photobionts during different stages of their life cycles.
• Another fungus growing on a lichen may share the alga of its host and later swap it for another species (e.g. *Diploschistes muscorum* growing on *Cladonia* spp.).

Some lichens contain both algae and cyanobacteria within the thallus. The main photobiont is normally the green alga, and cyanobacteria are localized in distinct structures (cephalodia). This has given rise to one of the greatest enigmas in lichen biology: that the same lichen can exist in several very

RIGHT *Trebouxia* sp., a unicellular green alga. The algae found most frequently in lichens belong to the genus *Trebouxia*.

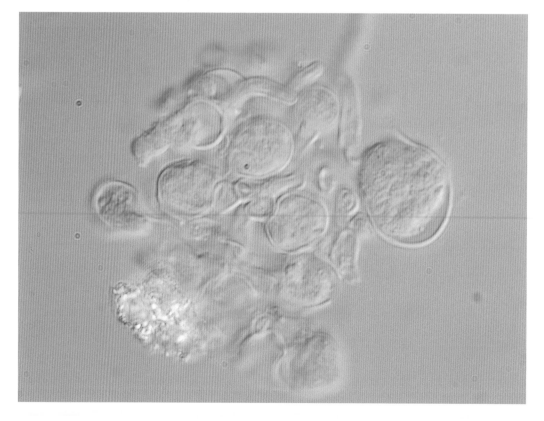

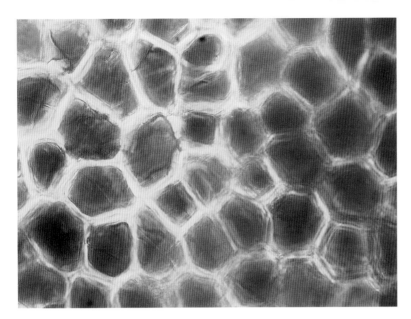

LEFT *Trentepohlia* sp. Green algae of this genus contain an orange carotenoid pigment, which masks the green colour of chlorophyll. Many lichens in shaded habitats and most lichens in lowland tropical rain forests contain *Trentepohlia*.

BELOW Chains of *Nostoc* sp. – a cyanobacterium. Members of this genus occur frequently in lichens, especially in those of wet habitats.

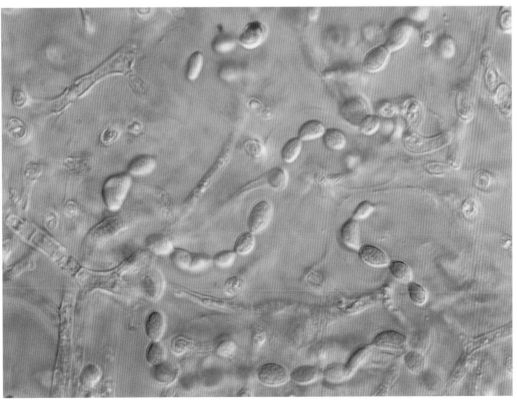

RIGHT *Sticta canariensis* – leaf-like form containing green algae, with large red-brown, disc-like fruiting bodies.

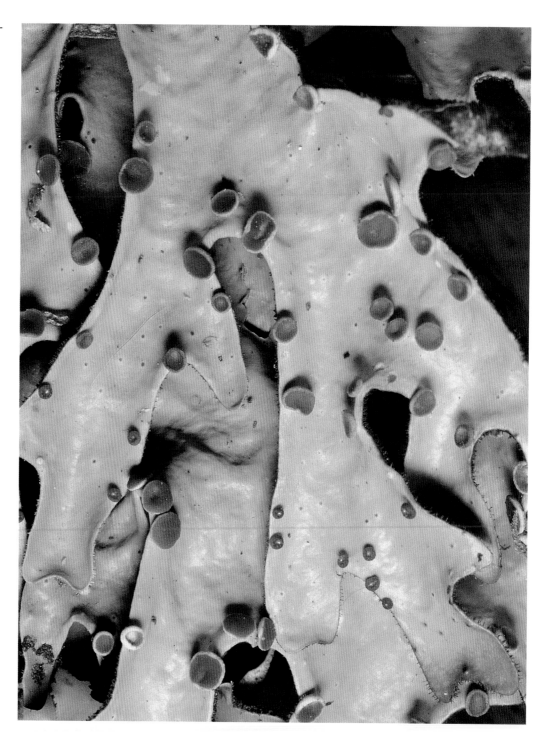

different forms, depending on the photobiont it contains. These different forms of the same lichen are called photomorphs.

# Photomorphs

One of the most important lichenological discoveries in the 20th century was that lichen algae play a major part in determining what a lichen looks like, how it functions and where it grows. Pioneering research was carried out by the Englishman, Peter James, and colleagues abroad. James discovered strange thalli in sheltered gorges in New Zealand and Scotland in which different lobes belonged to different species. For instance, in New Zealand he found the leafy lichen *Sticta felix* (containing green algae) growing out of the shrubby *Dendriscocaulon* species (containing cyanobacteria). He also found separate individuals of both lichens and realized that environmental factors (light, humidity) are, intriguingly, very important in determining which photobiont is present in the lichen. Because all lichen names are based on the fungus alone, a single name has to be used for what were previously considered to be two lichens. Many examples are now known, especially within the leaf-like lichens of the genera *Nephroma*, *Sticta* and *Lobaria*. In these groups, the same fungus may be associated with either a green alga and appear leaf-like or with a cyanobacterium, when it usually looks very different, often appearing shrub-like or as tiny granules. The different photobionts can either occur together in the same thallus or separately. Such an example is the lichen, *Sticta canariensis*. The variations are shown on pp.12-15; the lichen contains the same fungus but different photobionts. This particular example is from the cloud forest, Azores.

The differences in morphology, ecology, distribution and chemistry of the two independent lichens (photomorphs) can be so striking that at times they have been placed mistakenly in different genera when discovered unattached to each other. How do we know they are the same fungus? Early microscopical studies suggested this was the case because hyphae were observed to grow from one partner into the other.

BELOW **Zonation of photomorphs of *Sticta felix*, Lake Te Anau, South Island, New Zealand.**

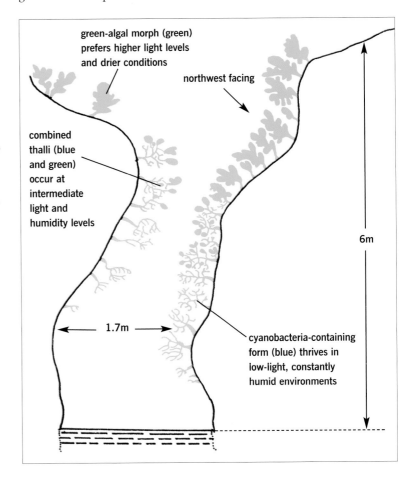

green-algal morph (green) prefers higher light levels and drier conditions

northwest facing

combined thalli (blue and green) occur at intermediate light and humidity levels

6m

1.7m

cyanobacteria-containing form (blue) thrives in low-light, constantly humid environments

LEFT *Sticta canariensis* – joined thalli. The small lobes (a) growing out of the thallus (b) contain green algae and the main thallus contains cyanobacteria.

This observation was later confirmed by experiments to resynthesize new lichens when the same fungus was inoculated with either a green alga or a cyanobacterium. More recently, molecular studies investigating DNA sequences have also confirmed that the same fungus is involved. An important and intriguing question is why do they look so different? We do not know the answer yet. In some ways the process of lichen formation appears to be similar to the formation of plant galls resulting from attack by insects or other small organisms. Different gall-causing organisms cause their host plants to form distinctive galls that are characteristic of the association.

## Structure

When we think about fungi we usually have a picture in our minds of mushrooms, toadstools, puffballs and other fungi that have conspicuous fruiting bodies, which usually occur only at certain times of the year and which, if edible, we can eat. But the fruiting body is only a small part of the whole fungus. A diffuse mycelium (a mass of feeding filaments called hyphae) spreads below ground, often for several metres or yards. Lichens are very different from non-lichenized fungi. The bulk of the lichen lies

TOP *Sticta canariensis* – leaf-like form containing cyanobacteria.

BOTTOM *Sticta canariensis* – shrub-like form containing cyanobacteria.

RIGHT **Section of layered leaf-like lichen** (*Parmelia caperata*): (a) outer cortex, (b) photobiont layer, (c) medulla and (d) melanized lower cortex.

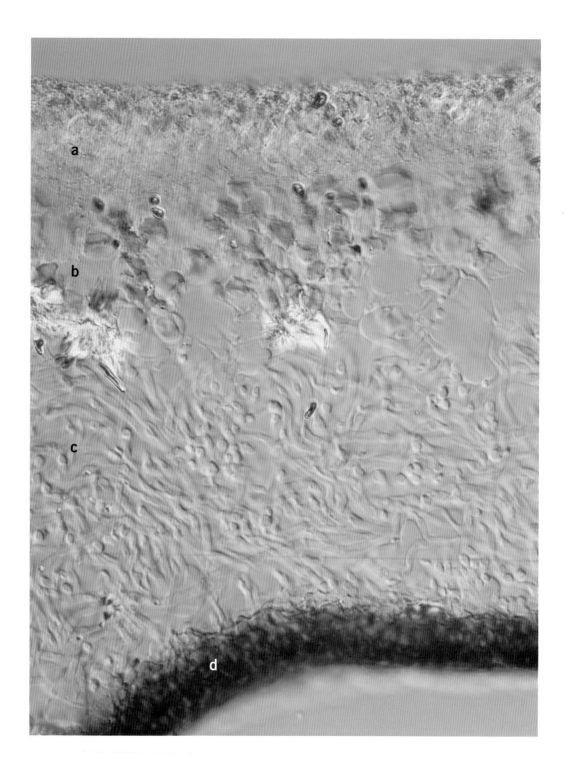

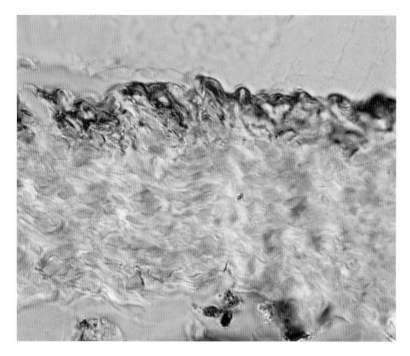

normally above ground, is visible throughout the year and the fruiting bodies are usually small compared with the thallus.

Fungi normally form the greater part of lichen thalli, the photobiont accounting for no more than 20%, often much less. Leafy lichens produce layered structures similar to leaves, consisting of an outer protective fungal layer (cortex) of closely compacted hyphae overlying a photobiont layer, which in turn overlies a loose medullary layer. A lower cortex is often present and the lower surface is usually attached to the substrate by root-like rhizines and other attachment organs. Other leaf-like lichens, such as the 'jelly lichens', which contain cyanobacteria, lack a layered structure and consist of intertwining mixtures of fungal and algal cells. Many beard-like lichens are concentrically layered with a hollow centre. The beard-moss lichens, *Usnea* spp., hang from trees and have central, tough, thread-like elastic strands to support their weight. Crustose lichens range from simple powdery crusts having little apparent structure to others with highly elaborate structures.

TOP **Section of non-layered leaf-like lichen (*Collema auriforme*). Chains of the photobiont (*Nostoc*) occur throughout the lichen (blue).**

BOTTOM **Section of shrubby lichen (*Usnea inflata*). The internal structure is radial: (a) dense central cord, (b) lax medulla, (c) thin photobiont layer, (d) dense outer cortex.**

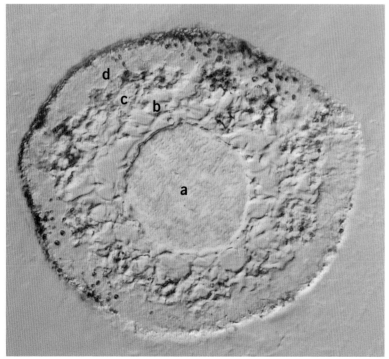

Lichen surfaces are often extraordinarily beautiful when studied with either a x10 magnifying lens or dissecting binocular microscope. Scanning electron microscopy reveals yet more detail. There is a wide range of structures and surface features, which help the lichen grow and reproduce. They are also useful in helping us identify and classify lichens. Both the upper and lower surfaces regularly have pores or cracks to allow gases to enter and leave the thallus. Pseudocyphellae are pores in the cortex where the loosely packed medulla bursts through to the surface. Pseudocyphellae occur in many different lichens but vary greatly in their shape.

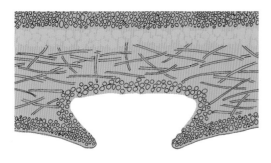

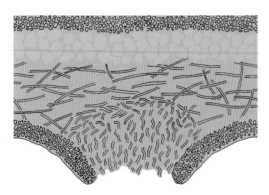

LEFT **Pores on the lower surface of *Sticta* sp. (cyphellae) (top), and *Psedocyphellaria* sp. (pseudocyphellae) (bottom).**

Cyphellae are neater, round, cup-like pores lined with globular cells restricted to the genus *Sticta*. Not all lichens possess cyphellae or pseudocyphellae but where they are present these hydrophobic structures are thought to act as pathways for gas diffusion.

## Amazing but true!

Some lichens have incredible powers to reorganize themselves and regenerate new tissues, for example when a lobe becomes turned over. The pioneering experimental Swiss lichenologist Professor Rosmarie Honegger carried out a simple, yet elegant experiment. She used a very strong glue to fix samples of *Xanthoria parietina* upside down to terracotta bricks, which were laid out on the flat roof of her institute. Honegger found that after 18 months in the field tiny new lobes developed at the edges of the fragments with the top surface correctly oriented in its normal position. Honegger suggested that the *Xanthoria* mycobiont can therefore sense and correct being upside down to enable the photobiont cells to receive adequate illumination. She also studied gold-coated samples under low vacuum conditions in a scanning electron microscope (SEM). Most researchers having completed their experiment would simply throw the samples away. But Honegger re-glued the now gold-coated samples on to the original bricks. She found to her surprise, that, provided the original samples were air-dried before she carried out her SEM study, they were capable of withstanding this harsh treatment and could continue growing.

# How lichens grow, multiply and disperse

## Reproduction and dispersal

Like other non-lichenized ascomycete fungi, most lichens produce sexual fruiting bodies containing spores. Unlike those of their free-living relatives the fruiting bodies are normally perennial, occur throughout the year and may be long-lived (eg. over 50 years in the Swiss Alps). Fruiting bodies in other lichens are seasonal and may be seen only at the cooler and moister times of the year. Hard to the touch and often difficult to cut without previous moistening, fruiting bodies may be flattened or semi-globose discs (apothecia), flask-shaped structures (perithecia), have stalks or resemble hieroglyphics. How sexual reproduction occurs in lichens is a mystery because it has never been observed directly under a

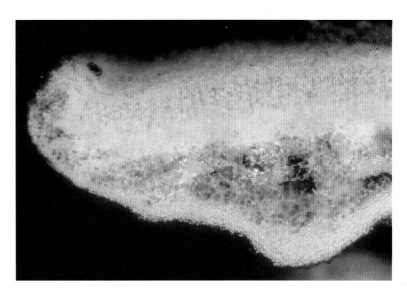

ABOVE **Section through apothecium (*Xanthoria parietina*). Algae are present in the margin of the fruiting body and beneath the spore-bearing layer. The upper surface of the apothecium is bright yellow due to the pigment parietin.**

LEFT **Section through an apothecium. There are two basic types, those lacking algae in the margin (left), and those containing algae in a thallus-like margin (far left). Eight ascospores are formed within the sac-like asci.**

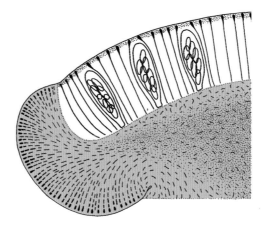

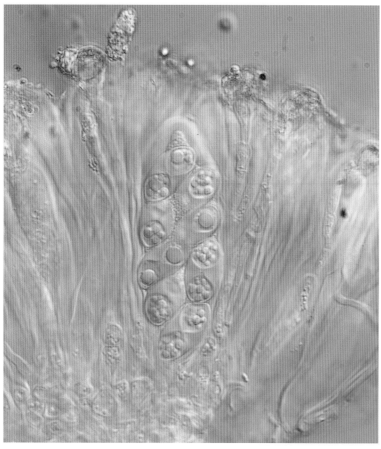

microscope. We believe that it is similar to sexual reproduction in other fungi and that fruiting bodies are developed initially from packets of specialized cells, which produce erect female filaments (trichogynes). These are fertilized by male cells (conidia) produced elsewhere on the lichen or possibly by other lichens of the same species growing close by. The nucleus of the fertilizing conidium joins with a female nucleus and they exchange genetic material by meiosis and divide by mitosis, producing ascospores. Ascospores vary in size, shape and structure and may be colourless or brown. Spore sizes range from 3 μm (0.003 mm) to over 250 μm (0.25 mm), and the larger sizes are visible under a x10 hand-lens or binocular microscope. Some are simple, without any internal divisions, while others have dividing cross walls. Normally eight spores are produced in the flask-shaped ascus, but some lichen asci may contain fewer – sometimes only one and sometimes several hundred. Spores are discharged either when the ascus wall layers split or, in the case of

ABOVE **Ascus containing eight ascospores (*Xanthoria parietina*).**

RIGHT **Structure of flask-like perithecium showing thick black outer wall and asci containing spores. Spores are ejected via the small opening (ostiole) at the top.**

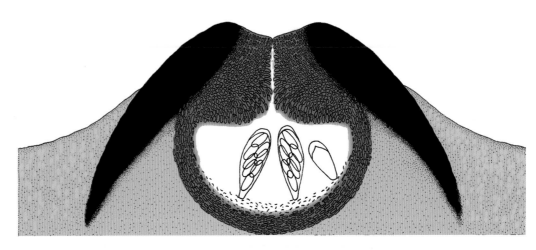

loose, dry spore masses, by rain-water droplets falling on them. On germination they need to meet a compatible alga to form a new lichen. Some lichens, such as *Staurothele* spp., conveniently carry algal cells in their fruiting bodies and eject these together with fungal spores, but this is rare among lichens.

In general, sexual reproduction normally ensures that populations retain genetic variability, which is important in maintaining the health and vigour of populations in the wild. But we know very little about the genetic variability of lichen populations. We do not know how often selfing (involving a single individual) or outcrossing (involving two individuals of the same species) occurs.

## Vegetative reproduction

Relying solely on reproduction by spores is a risky business because lichen fungi must find the right partner to form a new lichen. Many lichens increase their chances of success by reproducing vegetatively (asexually). They produce special propagules containing both fungus and photobiont. The most frequent are powdery structures (soredia), clusters of photobiont cells enveloped by fungal hyphae. Soredia may develop on the entire upper

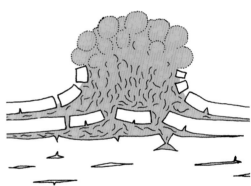

ABOVE **Diagram of structure of soralium containing soredia.**

LEFT **Close-up of powdery soralia (*Loxospora elatina*).**

surface of a lichen where the outer layer breaks down or in discrete patches (soralia), often depending on the species. Soralia differ enormously in their position on the thallus and their shape, and are important characters when identifying lichens. Another form of propagule, isidia, are smooth, peg-like outgrowths from the outer cortex and these also contain photobiont cells like the parent thallus. Some are simple, others branched like coral, or flattened. Soredia and isidia are scattered by wind, water and small animals, for example mites and birds, to new locations where, providing conditions are favourable, they grow into a new thallus.

Lichen communities on islands and also in formerly heavily polluted areas where lichens are reinvading, contain a greater proportion of species reproducing asexually, suggesting that this method of reproduction is very effective. Lichen fragments have been found on the feet of migratory birds and have been collected in high-altitude air samples from balloons above the Pacific Ocean. Some lichens are not known to reproduce sexually at all, as in the powdery lichens, *Lepraria* spp., and the worm lichen, *Thamnolia vermicularis* (see p. 69), which is found on the summits of mountains throughout the world and in the Arctic.

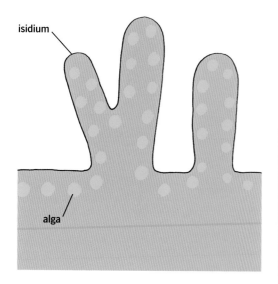

isidium

alga

ABOVE **Diagram of isidium structure.**

RIGHT **Close-up of isidia (*Pertusaria pseudocorallina*).**

## Mechanical hybrids

Sexual hybrids produced by genetic exchange between different species are unknown in lichens. However, lichen thalli may grow together and fuse. This may occur between populations of the same species, between different species of the same genus or between different genera, creating 'mechanical hybrids'. This is most easily observed where a pigment such as parietin (yellow) serves as a marker, as in *Xanthoria parietina*, which often occurs together with grey *Physcia* and related species.

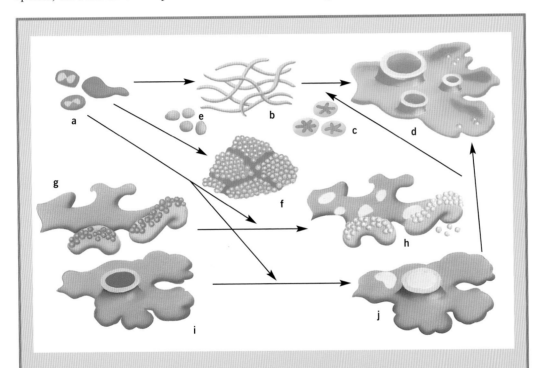

### Life cycle of *Xanthoria parietina*

(a) germinating fungal ascospores; (b) developing fungal hyphae; (c) free-living lichen algae; (d) mature *Xanthoria* formed by fungal hyphae meeting lichen algae; (e) foreign algae not involved in lichen formation; (f) undifferentiated 'lichen crust' containing fungal hyphae and foreign algae; (g) *Physcia* sp. producing powdery soredia consisting of fungal hyphae intermixed with lichen algae; (h) *Physcia* thallus and soredia infected by *Xanthoria* fungal spores or the undifferentiated crust in (f) resulting in *Physcia* disseminating *Xanthoria* through powdery soredia, which *Xanthoria* itself does not produce; (i) *Physcia* producing spore-bearing fruiting bodies – apothecia; (j) *Physcia* thallus and apothecia infected by *Xanthoria* resulting in *Physcia* thallus producing *Xanthoria* spores. Yellow colour represents tissues infected by *Xanthoria*.

## Species pairs

While some lichens have fruiting bodies and asexual propagules on the same thallus, others may exist in two separate forms, one in which sexual thalli produce fruiting bodies and spores, and another whose thalli reproduce asexually. These lichens are called species pairs. *Dirina massiliensis* (which is fertile) and *D. massiliensis* f. *sorediata* (which is asexual) are examples of this. They look very similar but the latter form is the most numerous. Analysis of DNA may help us resolve whether they are really a species pair. We believe, but have not proved it yet, that lichens reproducing asexually have evolved from the fertile species.

# Physiology

Lichens are able to colonize environments that have extremes of humidity, temperature and light, and they often occur in places where few other living things are able to survive. They have evolved a variety of ways to achieve this. One is their ability to switch off metabolic processes when they are dry and another is their limited need for nutrients. Lichen physiology varies according to the type of photobiont, lichen morphology and the presence of lichen substances.

Species pairs

TOP *Dirina massiliensis* (fertile).

BOTTOM *D. massiliensis* f. *sorediata* (asexual). The orange (scratched) surface is due to the pigmented alga *Trentepohlia*.

## The role of water

Some lichens rely on absorbing water from the atmosphere, for example from fog rather than from rain. Unlike flowering plants, lichens have no protective waxy outer cuticle, and therefore have little control over their water content. This fluctuates dramatically according to the environmental conditions. In dry conditions lichens contain up to 15–30% water and are metabolically inactive. This enables them to survive extreme periods of drought, perhaps lasting for several months. However, as soon as it rains or if there is a heavy dew, lichens

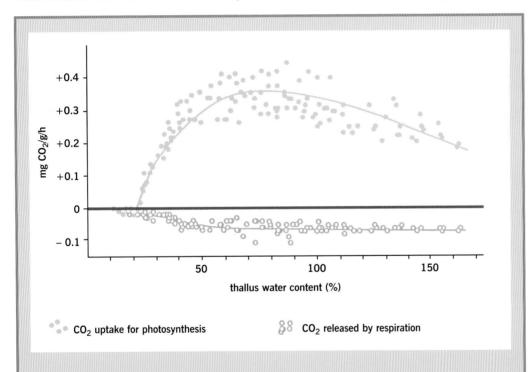

$CO_2$ uptake for photosynthesis          $CO_2$ released by respiration

## Effect of water content

A dry lichen is physiologically inactive and may absorb water from rain, dew or fog. Photosynthesis fixes carbon from carbon dioxide using chlorophyll and the energy of sunlight, whereas respiration produces carbon dioxide by breaking down sugars. Both processes occur in lichens and whether more carbon dioxide is fixed in photosynthesis than is released by respiration depends on how wet a lichen is.

The effect of water content on photosynthesis and respiration in the desert lichen *Ramalina maciformis* is shown above. At a water content of 22%, the amounts of carbon dioxide produced by photosynthesis and consumed in respiration are the same, so no net carbon is fixed as carbohydrate. Photosynthesis increases rapidly until a maximum is reached at 80% water content and then falls off at high water contents.

absorb water and can photosynthesize within minutes, provided there is sufficient light.

Net carbon fixation depends on the water content of the thallus, as well as on on the photobiont present. Lichens containing green algae typically absorb water to 2.5–4 times their oven-dry weight and those containing cyanobacteria can increase their weight by as much as 16–20 times. When they are dry, lichens are very brittle, can be crushed easily, and the photobionts are hard to see using a x10 hand-lens. When lichens are moist the surface becomes translucent and the algae can be seen easily. Some lichens look totally different when wet and may become bright green.

Lichens may tolerate extremes of temperature and drought depending on their environment, for example:
• Some may photosynthesize at -20°C (-4°F). They cannot survive at room temperature for more than a few months, but can remain viable at -20°C (-4°F) for at least 10 years, as for example in a domestic freezer.
• Dry lichens are metabolically inactive and are often unaffected by pollution. Lichens transplanted to cities in dry seasons normally appear healthy until it rains.

## Nutrition

The photobiont in a lichen produces carbohydrates through photosynthesis and the fungus absorbs these as sugar alcohols in the case of lichens that contain green algae, and as glucose in lichens that contain cyano-bacteria. The fungus then rapidly converts these products to the sugar alcohol, mannitol, for storage. In addition, cyanobacteria

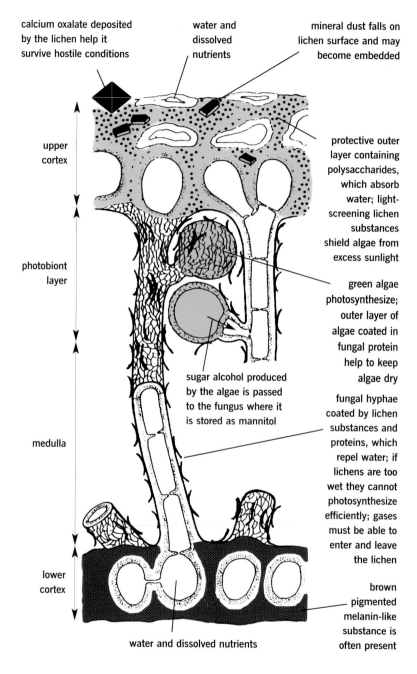

calcium oxalate deposited by the lichen help it survive hostile conditions

water and dissolved nutrients

mineral dust falls on lichen surface and may become embedded

upper cortex

protective outer layer containing polysaccharides, which absorb water; light-screening lichen substances shield algae from excess sunlight

photobiont layer

green algae photosynthesize; outer layer of algae coated in fungal protein help to keep algae dry

sugar alcohol produced by the algae is passed to the fungus where it is stored as mannitol

fungal hyphae coated by lichen substances and proteins, which repel water; if lichens are too wet they cannot photosynthesize efficiently; gases must be able to enter and leave the lichen

medulla

lower cortex

brown pigmented melanin-like substance is often present

water and dissolved nutrients

ABOVE **How water and nutrients are absorbed by lichens.**

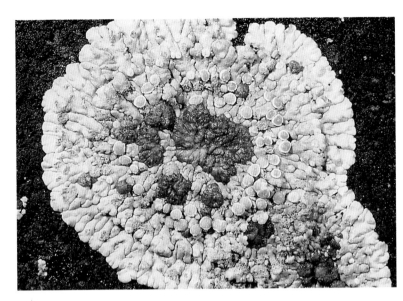

LEFT *Placopsis lambii*. The lichen is wet, which makes the colours appear brighter. The fruiting bodies are pale pink and the pale-brown cephalodia contain cyanobacteria. On coastal rock, Oregon, USA.

BELOW A characteristic assemblage of lichens on a rock that receives extra nutrients from bird droppings. It is dominated by the yellow lichen *Xanthoria parietina*. Oil refineries are visible in the distance. Pembroke, Wales, UK.

in lichens can also convert nitrogen from the air into ammonium ions ($NH_4^+$), which are available to the fungus for protein synthesis.

Lichens lack roots and normally grow in nutrient-poor habitats, so they rely on absorbing the limited nutrients and minerals they need from the atmosphere to sustain their relatively slow growth rate. This usually involves the whole surface of the lichen. Most nutrients are present in rain-water but some may be also obtained from the surface in run-off. In some places where birds roost or perch regularly, their accumulated droppings may be a source of additional nutrients. Here characteristic lichen communities occur, which include *Candelariella* spp., *Lecanora muralis* and *Xanthoria parietina*, lichens that contain the highest reported nitrogen contents (4.2 to 9.24%).

BELOW **Calcium oxalate crystals (SEM) on the surface of the lichen *Aspicilia mashiginensis*, which appears chalk-white. Gjersvik, Nord Trøndelag, Norway.**

# Lichen substances

Over 700 secondary lichen substances have been described from lichens, and new compounds are being discovered all the time. Many of these substances, which belong to chemically diverse classes of compounds – including aromatic compounds such as depsides, depsidones and carotenoids – are unique to lichen fungi. It was believed originally that these substances were produced only in intact lichens but, increasingly, we are finding lichen substances produced by mycobionts grown in pure culture. The presence of early biosynthetic products (precursors) in mycobiont cultures suggests that the photobiont may also play a role in determining which chemical products are produced by a lichen.

Many substances affect the appearance of lichens. In the pruinose lichen *Physcia aipolia*, the surface of the fruiting body appears bluish-grey as a result of a thin layer of calcium oxalate crystals, while crystals of the same substance on the surface of *Aspicilia mashiginensis* make it look chalk-white. This substance is especially frequent in lichens growing on limestones and also in arid habitats. Oxalates may help deflect the amount of light reaching a lichen and may help lichens survive in extreme environments.

Lichen compounds may occur in extremely high concentrations, occasionally up to 20% of the oven-dry weight of the lichen. This is a lot, given that lichens grow so slowly. We do not know why most of these substances are produced but it is hard to believe they are all simply waste

LEFT **A pruinose lichen (*Physcia aipolia*). A thin layer of calcium oxalate crystals gives a bluish-grey appearance to the surface of the fruiting body. On aspen, *Populus tremula*, Hälsingland, Sweden.**

products. Some popular ideas include:

• The bitter taste of some lichen substances may make the lichen unpalatable to grazing animals, such as slugs and snails.

• Some substances may be important as 'stress metabolites' for survival. For instance, many lichen species are more brightly coloured in sunny sites and contain higher levels of secondary metabolites than when growing in shadier habitats, suggesting protection against ultraviolet (UV) radiation. For example, *Xanthoria parietina* is bright yellow in sunny habitats and grey in the shade.

• Some lichen substances have known antibiotic activity and may inhibit growth of faster growing plant species.

• Certain lichen acids may be important in detoxifying toxic metals.

• Lichen acids (especially oxalic acid) may help to dissolve insoluble essential minerals necessary for lichen growth.

• Hydrophobic (non-wettable) lichen substances may improve gaseous exchange, for example in the medulla. This is particularly important for lichens living in wet habitats because they cannot photosynthesize if they are too wet.

OPPOSITE **The wolf lichen,** *Letharia vulpina*, **is brightly coloured because it contains the lichen substance and metabolic poison vulpinic acid. Formerly used in Scandinavia to kill wolves and foxes. British Columbia, Canada.**

RIGHT **Thread-like crystals of salazinic acid coating hyphae in the medulla of** *Parmelia sulcata*.

## Identifying lichens from the substances they contain

From the very earliest studies colour has been used to identify lichens. The genus *Physcia* is similar in appearance to *Xanthoria* except that *Physcia* (containing the colourless substance atranorin) is grey and *Xanthoria* (containing an orange pigment, parietin) is yellow-orange. Most secondary substances are colourless, however, and chemical analysis is required to detect and identify them.

There has been a long history of using chemical tests to identify and classify lichens. This dates back to the simple spot-test reactions first performed in the 1860s by Finnish-born William Nylander (who emigrated to live in Paris) using potassium hydroxide solution and bleach. Later the pioneering Japanese Yasuhiko Asahina, who determined the structures of many lichen chemicals, also developed an additional spot reagent (P or PD, an alcoholic solution of *p*-phenylenediamine), which reacted with aromatic aldehydes to become orange or red. Subsequently a wide range of analytical methods including microcrystallization, Thin Layer Chromatography (TLC) and other sophisticated techniques were developed. The study of 'chemosystematics' was pioneered by lichenologists.

The diversity of lichen substances provides an incredibly useful chemical fingerprint to help identify and classify lichens. Indeed, there is little alternative in the case of many sterile crustose lichens or for those that possess few diagnostic characters. Most lichens can be identified using relatively simple spot-test reactions and chromatography.

## Tests for lichen substances

Spot-test reactions of lichens (right). *Xanthoria* contains the yellow pigment, parietin, which changes to a deep purple-red when a drop of caustic potash (potassium hydroxide solution) is placed on it. When the same chemical is applied to *Physcia* (grey), which contains the colourless pigment, atranorin, there is no significant colour change.

Microcrystallization tests (bottom left) involve extracting lichen substances in acetone, evaporating the solvent and recrystallizing the residue in a suitable solvent on a microscope slide. Different substances crystallize in distinctive shapes and colours. A simple technique, but now largely superseded by chromatography and other modern techniques.

Thin Layer Chromatography (TLC) plate (bottom right). Lichen substances extracted in acetone, applied to silica-gel-coated TLC plates and run in standard solvents provide a convenient method of identifying lichen substances. Different substances travel at different rates in solvents and produce characteristic colours when viewed in either natural daylight or under UV light.

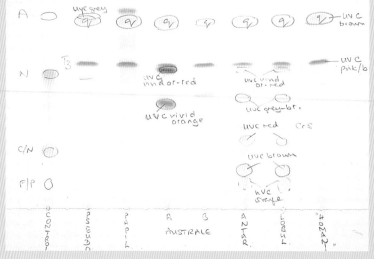

# Lichen biodiversity

Biodiversity is the variety of life. It includes all organisms and the diversity within species, between species and of whole ecosystems. The Convention on Biological Diversity, signed in Rio de Janiero in 1992, highlighted the importance of "the conservation of biological diversity, the sustainable use of its components and the fair and equitable sharing of the benefits arising out of the utilization of genetic resources." The Convention is the first global, comprehensive agreement addressing all aspects of biological diversity: genetic resources, species and ecosystems. It recognizes, for the first time, that conserving biological diversity is important and of concern to humankind.

No one knows for certain how many lichen species there are on Earth – there is no comprehensive global list. Estimates vary from a conservative 13,500 to over 30,000. It is not difficult to discover new records and species of lichens even in comparatively well studied areas. The British lichen flora is probably the best known in the world but over 100 new lichens, including several new to science, have been discovered in Britain since 1992. Numerous crustose lichens have been overlooked or else 'lumped' together with other species and only recognized as being distinct following recent detailed taxonomic research. In the rain forests of British Columbia, 20 species are added to the list every year. The extent of genetic variation among lichenized fungi at species and population levels, and also of lichen photobionts, has hardly been investigated. Initial studies in areas where lichens have been depleted through former air pollution or forestry practices and have since reinvaded, suggest that less genetic diversity may exist among these species than before. Lichens also have an extraordinary chemical diversity and new compounds are being discovered continually.

Many human activities create important new habitats for lichens, which would otherwise be very rare or absent. For instance, in areas where natural rock exposures are scarce or absent, as in South-east England, over 300 lichens may now be found colonizing gravestones in churchyards. Mine-spoil heaps and even abandoned aircraft runways provide rich areas for lichen biodiversity.

BELOW *Psilolechia leprosa*, an overlooked lichen. Described as new in 1987, it occurs throughout Britain, mainly on church walls by copper lightning conductors. Also recorded from Greenland, Norway and Sweden from old mine sites.

Much of the interest in lichens started in Europe and North America and this has spread to Argentina and Chile, parts of Brazil, Japan, Australia and New Zealand. Lichenology is now developing in South-east Asia, Africa and other areas of South America as more and more countries are recognizing the importance of lichens, particularly for monitoring environmental change. In the next few pages we will take a brief look at the amazing diversity of lichens and discuss their evolution.

## Mushroom lichens

Everyone is familiar with the common toadstool or mushroom and puffballs. Most belong to the basidiomycetes, which produce their spores on the outside of a special cell called a basidium, rather than within the sac-like ascus as in the sac fungi (ascomycetes). Very few lichenized basidiomycetes occur. There are basically three different types, classified according to the type of their fruiting bodies. One type is mushroom-like with gills, for example *Omphalina umbellifera*. They

BELOW **Mushroom-shaped** ***Omphalina umbellifera*,** **growing on a log of** ***Nothofagus cunninghamii*** **in a rain-forest clearing, Tasmania, Australia.**

produce soft fruiting bodies, identical to those of other mushrooms, which develop from mats of algae and appear like granules or small scales. These lichens are found throughout the world in damp habitats on soil, rotting wood, among *Sphagnum* mosses or on other lichens. The second type, for example the club-shaped *Multiclavula vernalis*, produces simple pale flesh-coloured, orange or white clubs from mats of algae in peaty soil in mountainous areas in the Northern and Southern Hemisphere. Finally, the bracket-shaped lichens, for example *Dictyonema glabratum,* which contain cyanobacteria, are widespread in submontane tropical rain forests.

LEFT **Club-shaped** *Multiclavula vernalis* **Tasmania, Australia.**

BELOW **Bracket-shaped** *Dictyonema glabratum* **Chiriqui, Panama.**

# Coral lichens and pin lichens

The coral lichens and pin lichens include several unrelated lichens, which all have a passive rather than an explosive mechanism for spore dispersal. The simple spore-sacs disintegrate when the spores are still young. The spores build up as a dry thick layer rather like soot (mazaedium) where they grow and mature. As in many other fungi with passive spore dispersal, spore ornamentations are often present, which aid dispersal by insects. Many of the crust lichens of this biological group have stalked, pin-like, fruiting bodies, but some large shrubby lichens also form mazaedia.

Coral lichens are big shrubby lichens, which favour cool, moist habitats. Most species occur in the temperate rain forests of the Southern Hemisphere but three are found in Britain. The globose fruiting bodies produce a prominent sooty mazaedium. The genus *Sphaerophorus*, which derives its name from the spherical fruiting bodies, include some of the most beautiful and distinct of all lichens.

Pin lichens are much smaller and require a x10 hand lens to be seen properly. Species within the genus *Calicium*, which are indicators of undisturbed old-growth forests in Northern Europe, often have beautifully ornamented spores, which are important for identification. *Chaenotheca ferruginea* is a common, pollution-tolerant pin lichen, which can be found in parks and forests close to urban and industrialized areas. Easily recognizable by the rust-coloured spots on its thallus, it may form extensive patches on sheltered trunks.

All lichens producing these specialized mazaedia fruiting bodies were, until recently, regarded as being closely related, but recent DNA analyses carried out by the Swedish lichenologist Dr Mats Wedin have revealed that they are derived from a range of fungal groups. The pin-lichen genera *Calicium* and *Chaenotheca*, although appearing very similar, are not related.

BOTTOM LEFT
*Sphaerophorus globosus* – a coral lichen.

BOTTOM RIGHT *Calicium quercinum* – a pin lichen.

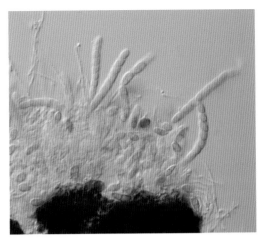

LEFT *Calicium denigratum* showing asci and part of a sooty mazaedium.

RIGHT *Chaenotheca ferruginea*, a common, pollution-tolerant pin lichen.

# *Parmelia* lichens

*Parmelia* lichens form the largest group of leaf-like lichens, including over 1000 species found on rocks, bark, wood and soil throughout the world. The thalli have an upper and lower cortex and are normally loosely attached to the substrate by rhizines. The colour of the upper surface ranges from grey (atranorin), to yellowish (usnic acid) or brown (melanin-like pigments), according to which substances are present. This is a group every beginner quickly learns to recognize.

Nowadays this unnatural group has been reorganized into many smaller genera according to a range of fungal morphological and chemical characters.

Their abundance has led to their use in numerous monitoring studies, including estimating growth rates and assessing heavy-metal contaminants. Several studies using different techniques suggest that individual *Parmelia* lichens may not be genetically identical. Thalli of the same species often fuse together when they meet, even if they are

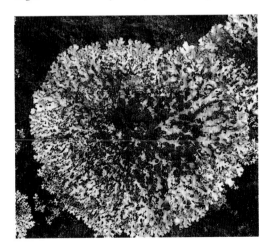

LEFT **Examining the parmelia lichen** *Hypotrachyna britannica* on Flores, Azores, at its only known site in the archipelago. Is this a recent colonist or is it rare because of a lack of suitable substrates?

RIGHT *Hypotrachyna britannica*, showing the dark indigo-blue soralia in the centre of the thallus.

genetically different strains. Parts of *Parmelia* thalli have been shown to be chemically distinct from one another and, at times, different lobes within a thallus may grow at different rates. A loss of senescent inner thallus areas is common in *Parmelia* lichens and many other leafy macrolichens and may result, as in *Arctoparmelia centrifuga*, in the formation of ring-shaped, zoned thalli. Bare patches of the substrate may be recolonized by younger stages arising from asexual propagules (soredia, isidia).

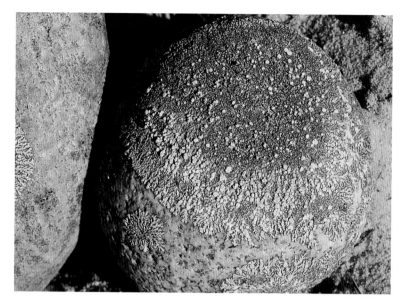

LEFT *Xanthoparmelia mougeotii* on a granite boulder, Cuillin Forest, Scotland, UK, showing the yellowish powdery soralia. This lichen is spreading in Southern Britain on man-made structures.

LEFT *Arctoparmelia centrifuga* and other lichens on boulders near Reserve Faunique, Laurentides, Quebec, Canada.

# Reindeer moss and pixie-cups

Reindeer moss and pixie-cup lichens
(*Cladonia* spp.) are major components of
lichen-rich heathlands worldwide. The
fruiting bodies are branched or cup-like and
grow from mats of horizontal thalli. Reindeer
moss lichens, which may cover the ground
like a thick hoar frost, often resemble small
shrubs and architects use them for this
purpose in their models. They propagate
themselves mainly by fragmentation,
although most species occasionally form
spores. Many species contain a UV-absorbing
compound in the outer region, either
yellowish usnic acid or the colourless
atranorin. These pigments are responsible for

LEFT **Reindeer moss
lichen, *Cladonia stellaris*,
with strands of green
club moss, *Lycopodium
clavatum*. Northern
British Columbia,
Canada.**

LEFT ***Cladonia cristatella***
**(British soldiers), with
a stalk of *Cladonia
deformis* on the lower
right. White Mountains,
New Hampshire, USA.**

LEFT **Close-up of reindeer moss lichen,** *Cladonia rangiferina*. **Cairngorm Mountain, Scotland, UK.**

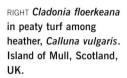

RIGHT *Cladonia floerkeana* in peaty turf among heather, *Calluna vulgaris*. Island of Mull, Scotland, UK.

BELOW *Cladonia chlorophaea*. On earth bank, Deeside, Scotland, UK.

the yellowish or greyish coloured mosaics so characteristic of lichen carpets in heathlands and northern forests. Reindeer moss lichens lack an outer cortex and never form cups. Pixie-cup lichens are closely related to the reindeer moss lichens. In some species, the fruits are a striking scarlet, due to the pigment rhodocladonic acid; others have brown or pale-coloured fruits.

*Cladonia* is a large genus containing several hundred species and is probably the first group of lichens beginners learn to recognize. They are often morphologically variable, with subtle differences in their

branching patterns, shape of cups, size of soredia coating cups, or lack of cups. They may be difficult to name to species level, especially for specimens lacking cups, without first carrying out chemical studies.

## Script lichens

Script lichens are crustose lichens, which have simple stretched or branched fruiting bodies that often conceal the central spore-bearing surface. The outer fruit-body wall is often thick, jet black and rigid and the fruits may

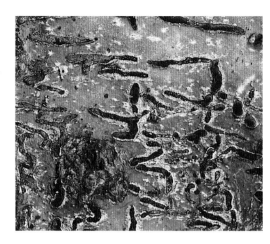

RIGHT *Phaeographis lyellii*, on young oak, *Quercus robur*. Dorset, UK.

LEFT Hazel coppice with oak standards supporting several script and other crustose lichens. Powerstock Common, Dorset, UK.

RIGHT *Opegrapha filicina* growing on a leaf. Costa Rica.

BELOW Section of *Phaeographis dendritica* showing black outer fruit-body wall and fruit itself.

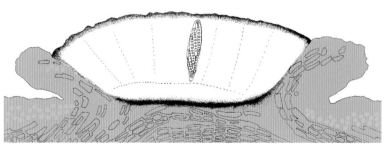

resemble strange hieroglyphics. Script lichens often occur on smooth bark in moist, shaded habitats. They contain algae of the genus *Trentepohlia* and appear orange when scratched. They are dominant in lowland tropical forests, covering tree trunks and may also grow on leaves.

## Flask lichens and pyrenolichens

Flask lichens and pyrenolichens are inconspicuous and the thallus may be entirely immersed within the rock or bark substrate. The characteristic fruiting bodies (perithecia) look like miniature bottles and are often partly buried in the substrate. The outer wall is usually black, very hard and brittle. Spores are released from the perithecia through a minute opening at the tip, which is visible with a x10 hand-lens.

RIGHT *Parmentaria chinensis* (a) and *Pyrenula dermatodes* (b) on holly, *Ilex azorica*, Azores. Trees with smooth bark are much better for pyreno lichens than those with rough bark.

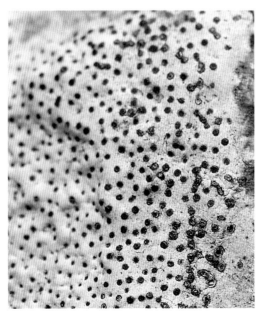

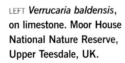

LEFT *Verrucaria baldensis*, on limestone. Moor House National Nature Reserve, Upper Teesdale, UK.

RIGHT *Pyrenula macrospora* on smooth bark of hazel, *Corylus avellana*, Western Scotland.

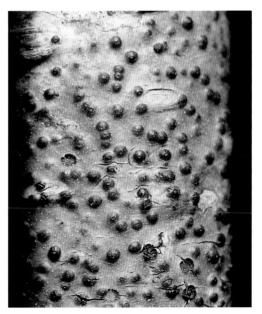

43

## Other crust and filamentous lichens

*Cryptothecia rubrocincta*, or Christmas lichen, is a widely distributed tropical and subtropical species occurring in South-eastern USA. *Cryptothecia* is unusual because discrete fruiting bodies are not formed and asci arise directly from the medulla. Asci have not yet been observed in this particular species and we do not really understand where it belongs systematically. It contains the red pigment, chiodectonic acid. This specimen is from northern Florida.

*Lecanora muralis* is one of the most successful urban lichens in the world and is found on roofs, walls and tarmacadam pavements. Its appearance is variable according to its environment. A study carried out by Professor Mark Seaward, University of Bradford, UK, between 1970 and 1980 found that *L. muralis* spread towards the centre of Leeds at an average rate of 150 m (164 yd) a year by colonizing asbestos-cement roofs following improvements in air quality. It is possible that an 'urban super-race' may be present. This lichen is often mistaken for chewing-gum.

*Psilolechia lucida* has a powdery, bright yellow-green thallus, often coating acid gravestones in Britain where the lurid colour looks like paint. Often sterile, this specimen is fertile and was collected from a Swedish barn painted with the characteristic red iron-oxide-based paint. The bright colour is due to rhizocarpic acid. In polluted environments *P. lucida* is rarely fertile.

*Caloplaca flavescens*. *Caloplaca* species are similar to the leaf-like *Xanthoria* but are firmly attached to the surface and lack a lower cortex. A study carried out in churchyards in London

ABOVE **Cryptothecia rubrocincta**

discovered that *C. flavescens* was restricted to pre-19th-century gravestones and appeared unable to colonize more recent stones. This suggested that the establishment phase was more sensitive to sulphur dioxide pollution than the mature phase.

*Arthonia tavaresii* belongs to a large group of 'lichenicolous fungi' or fungi that grow on lichens, in this case *Pyrenula*. The fruiting bodies are coated in a blood-red pigment. This species is restricted to the cloud forests of the Azores and the Canary Islands. *Arthonia* is one of the largest genera of crustose lichens, comprising more than 500 species as currently defined.

*Ramonia azorica* is restricted to a single endemic tree, *Juniperus brevifolius*, in the Azores, where it overgrows mosses. Its survival depends on the survival of relict laurel forests dating back to the Tertiary period. In these persistently wet rain forests, mosses are more abundant than lichens. This is one of a few lichens that can actually overgrow and kill mosses.

*Coenogonium leprieurii* is a filamentous lichen containing the green alga *Trebouxia*. *Coenogonium* is one of a few lichens where the alga is the dominant partner, and it is restricted to the low light environments of tropical and subtropical forests. This specimen is from Costa Rica.

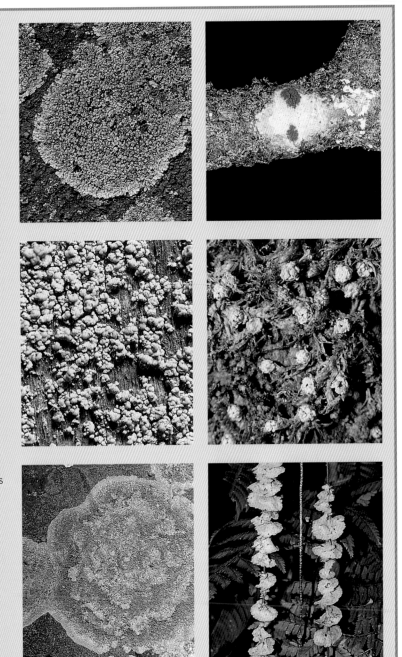

LEFT COLUMN

TOP **Lecanora muralis.**

MIDDLE **Psilolechia lucida.**

BOTTOM **Caloplaca flavescens.**

RIGHT COLUMN

TOP **Arthonia tavaresii.**

MIDDLE **Ramonia azorica.**

BOTTOM **Coenogonium leprieurii.**

# Evolution, classification and naming

## Evolution

Lichens or lichen-like associations are often considered to be among the first living things that emerged from the primeval soup and enabled higher organisms to conquer land. The lichen-habit may have allowed both fungal and algal components to exploit new ecological niches and to contribute to early soil formation on Earth.

Normally, fossils tell us about the early history of a group, but neither lichenized nor non-lichenized fungi are well preserved in the fossil record, so we know little about the earliest lichens. Relatively few people are actively looking for fossil lichens, which also explains our lack of knowledge. Filamentous microfossils estimated to be over 2 billion years old were once considered to be early lichens but nowadays this appears unlikely. Thick-walled spores and conidia, which may be fungal, have been found in Cambrian rocks (c. 570 million years old). However, the oldest undisputed fossil lichen, *Winfrenatia reticulata*, is from the Lower Devonian (c. 400 million years) Rhynie chert formation in Scotland. This association between fungal hyphae and a cyanobacterium forms a simple thallus with pockets containing possible soredia-like structures. The oldest fossil ascomycete fungus was also found at the same location.

The lichen lifestyle is widespread among very different fungal groups. It has clearly evolved independently numerous times (polyphyletic), and early DNA investigations of lichens also confirmed the long

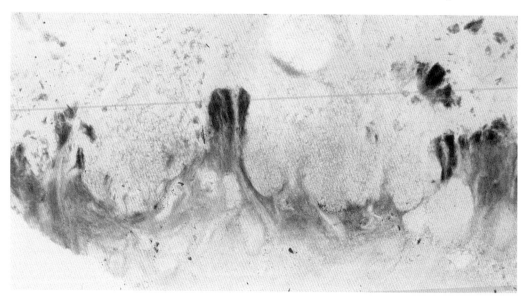

LEFT **400-million-year-old fossil lichen, *Winfrenatia reticulata*, preserved in the Rhynie Chert, Aberdeen, Scotland, UK. The thallus consists of layers of hyphae, which lack cross walls. The uniform, soralia-like depressions on the upper surface have a three-dimensional hyphal net-like structure extending into their base.**

acknowledged idea that lichens are polyphyletic in origin. In the fungal group basidiomycetes, lichens may have arisen at least three times. In the ascomycetes, at least four independent origins, probably more, can be predicted. In some crustose groups, repeated lichenization, de-lichenization and re-lichenization events seem to have taken place throughout evolution.

Interestingly, recent molecular studies provide no support for the idea that lichens living today are 'ancient' compared with other fungi. In fact, Lecanorales, the order to which most lichens belong, is likely to be a relatively advanced group of ascomycetes. Of course, this evidence does not mean that lichen-like associations were not among the first to conquer land, but if they were, present-day lichens are not likely to be their descendants. Analyses of DNA sequences and other techniques from molecular biology have caused a surge of investigations into the

evolution of lichenized fungi. These investigations have already contributed significantly to our understanding of the natural relationship of many lichen groups. Present-day lichen classifications will certainly change dramatically in the near future as we make progress in this field.

## Classification and naming

Humans have long needed to name and classify organisms to be able to grasp the overwhelming diversity of life. Lichens are no exception. Systematics, the scientific study of the diversity of life, is a very wide discipline including identification, naming, evolutionary studies and classification. Taxonomy is the description of taxa and taxonomic units (species, genera, families etc.). A scientific species name consists of two Latin or Greek words: the generic name and the species name. This binomial nomenclature,

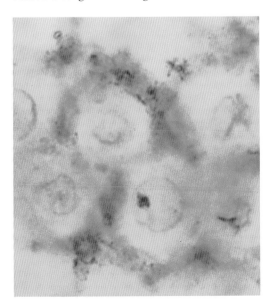

LEFT **A hyphal net surrounds a round cyanobacterium.**

RIGHT **Diagram showing hyphae enveloping photobiont cells near the base of the wall of a soralia-like depression.**

introduced by the Swedish naturalist Carl Linnaeus in the 18th century, rapidly replaced the old, cumbersome system of long descriptive phrase-names. Phylogeny refers to the tree-like, hierarchical pattern of evolutionary relationships between organisms. Classification is the arrangement of named taxa in a hierarchically structured system. Modern classifications aim to reflect the natural relationship between organisms.

The name we give to a lichen refers to the fungal partner; lichen algae have separate names. In the mid-19th century Simon Schwendener suggested that lichens should be classified among the fungi but lichenologists and mycologists long refused to accept his discovery, and continued to place lichens in a separate taxonomic group. It was not until the 1950s that the integration of lichens into

fungal classification became accepted more widely, and still today lichenized and non-lichenized fungi are usually studied by different researchers at different institutions. Almost all scientists today, however, agree on integrating lichenized fungi into the fungal classification.

Morphology and anatomy of the fungal fruiting bodies provide most characters that have been used in higher-level classifications of lichen fungi. Variations in the apical parts of the spore sacs, as indicated by their staining by iodine, have been used widely to delimit lichen families. Spores vary immensely in shape, size and colour, and have been used extensively when describing genera and species. At present, lichen systematics is at an exciting stage of development. Many larger 'genera' are being broken down into smaller genera – spore characters are no longer the only criterion for generic separation. Recent advances in molecular techniques, in combination with morphological, anatomical and chemical studies, are providing valuable tools for understanding lichen phylogeny and classifications.

LEFT **Erik Acharius, 1757–1819. The 'father of lichenology' and founder of modern lichen taxonomy.**

### Classification system

Kingdom: *Fungi*

Phylum (Division): *Ascomycota*

Subphylum (Subdivision): *Pezizomycotina*

Class: *Lecanoromycetes*

Family: *Lecanoraceae*

Genus: *Lecanora*

Species: *campestris*

Subspecies: *dolomitica*

# Ecological role

Vegetation dominated by lichens covers around 8% of the Earth's land surface. Lichens are dominant throughout the arctic tundra covering many thousands of square kilometres. Their dominance means they have a globally important role to play in plant ecology. They act as carbon sinks by consuming carbon dioxide used in photosynthesis and therefore play a part in delaying global warming. Where lichens cover the ground they prevent soils from drying out. Their ability to capture fog and dew is important in conserving moisture where water is scarce, as in deserts and also sheltered underhangs where they are not normally wetted directly by rain. On nutrient-poor soils in northern forests, they accumulate and release nutrients (nitrogen and phosphorus) required by forest trees for growth. In these forests a combination of fires and allelopathic lichen substances (which inhibit the growth of some tree seedlings), help to create large gaps in spruce woodlands in subantarctic areas of Canada and Russia. Nitrogen leached from lichens or fixed by lichens containing cyanobacteria is important for tree growth. Inputs for these regions vary

LEFT **Northern subantarctic woodland, Canada, carpeted by shrubby reindeer lichen, *Cladonia*, and *Stereocaulon* spp.**

between 1 and 40 kg (2.2 and 88 lb) of nitrogen per hectare per year.

Lichens are an important energy source for many animals. Tiny invertebrates as well as larger vertebrates eat them, including reindeer (called caribou in North America), *Rangifer tarandus*, black-tailed deer, *Odocoileus hemionus* and Chinese musk deer, *Moschus moschiferus*. In North America the spruce grouse, *Canachites canadensis* and wild turkey, *Meleagris gallopavo*, feed on them. Birds use them to build their nests. For example, in Madagascar the olive-headed weaver, *Ploceus olivaceiceps*, makes its nest entirely from *Usnea* spp. and in Europe the goldfinch, *Carduelis spinus*, builds its nest mainly from *Usnea* spp. Northern flying squirrels, *Glaucomys sabrinus*, use the beard-lichen, *Bryoria fremontii*, as nesting material and food. Some animals and insects use lichens for camouflage, including several moths, such as the peppered moth, *Biston betularia,* and grey dagger moth, *Acronicta psi*, in Europe, and certain butterflies store lichen compounds in their tissues as a chemical defence. Problems can arise through the ability of lichens to accumulate pollutants, including toxic metals and radionuclides. If contaminated lichens are

BELOW LEFT **Goldfinch, *Carduelis carduelis*, nest built with lichens, principally *Usnea articulata*. Cornwall, UK.**

BELOW RIGHT **Birds such as the royal albatross, *Diomedea epomophora*, may transport lichen propagules vast distances. Auckland Islands, New Zealand.**

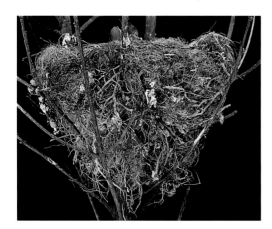

eaten the pollutants may be transported through the food chain to humans.

Lichens occur from below tide level on rocky coasts to near the summits of the highest mountains from the North to the South Pole. They are found on rocks, trees, the ground, mosses, other lichens and occasionally on leaves, the only major habitat where they are absent being the deep sea. Many lichens have distinct habitat preferences and ecological requirements. The physical and chemical properties (including pH) of the substrate are very important in determining which lichens develop. Some 'weedy' species may grow on virtually any substrate – on different trees, rocks, dustbin lids and rusting car bodies – while others are confined to individual tree species or rock types. Characteristic lichen communities develop according to a range of factors, including climate and substrate.

## Rock weathering

Lichens play an important role in breaking down rock minerals both physically and chemically and contribute to soil formation. Physical weathering occurs mechanically through hyphae and rhizines simply growing

LEFT **The grey dagger moth,** *Acronicta psi*, **camouflaged by lichens. North Wales, UK.**

ABOVE **Dunite, a rock composed of the silicate, olivine, colonized by *Lecidea lactea*. The lichen secretes oxalic acid and breaks down the rock, producing a characteristic dark-brown, iron-rich stain.**

LEFT **Scanning electron micrograph of the iron-rich stain beneath *Lecidea lactea* showing amorphous silica coated in iron produced by lichen weathering.**

through the rock resulting in the breaking up of minerals. This may seem an enormous task for such small organisms. But just consider the amazing ability of mushrooms to push up asphalt and concrete in driveways. Fungal hyphae are able to exert immense turgor pressure, even when they measure only a few microns in diameter. Lichen hyphae may penetrate minerals, especially those with well-developed cleavage planes, such as biotite (a ferro-magnesian mica). We do not know how much weathering is the result of lichen activity. Some people think that lichens may sometimes protect surfaces from other kinds of weathering (frost action, wind abrasion and acidic solutions produced by atmospheric pollutants) by binding together minerals with their hyphae.

Lichens may alter rocks chemically in three ways:

1. Simple organic acids, such as oxalic acid secreted by lichen fungi, may react with metals to form metal oxalates. By far the most abundant are the calcium oxalates, weddellite ($CaC_2O_4.2\text{-}3H_2O$) and whewhellite ($CaC_2O_4.H_2O$), which are frequent in and on lichens growing on calcareous rocks rich in calcium carbonate, such as limestones. Some lichens, such as *Dirina massiliensis* f. *sorediata*, may form thalli up to 1 cm (0.4 in) thick and appear chalk-like in cross-section because they contain such high concentrations of calcium oxalate. Sometimes it is difficult to identify the boundary between a lichen and its mineral substrate.

2. The outer layers of hyphae in contact with mineral grains are coated in acidic polysaccharides (sugars). These may absorb

water and assist in mineral breakdown by extracting particular metal ions from mineral surfaces.

3. Some lichen substances, such as depsides, are not very soluble in water and, in laboratory experiments, have been found to remove metals from minerals to form metal complexes. We do not know how often this occurs in nature.

Lichens colonize ancient monuments and building stones but do not usually cause significant damage. Architects and many other people recognize that lichens can give a desirable rustic charm to buildings, adding to local character. If you spray dilute manure on surfaces you can encourage growth of the colourful yellow *Xanthoria* spp. Lichens can be removed where necessary by using fungicides, bleach and caustic washes.

## Living rocks

Lichens cover rock surfaces throughout temperate and arctic regions of the world. In fact, in rocky landscapes you often cannot see the rocks themselves because of the lichens that clothe them. Indeed, geologists have been dismayed by the colonization of rocks by lichens because valuable geological information is obscured. Today, however, earth scientists recognize that lichens may provide important information – for instance in field mapping studies and remote sensing – because lichen communities on different types of rocks look different, even to the untrained eye. The colourful patina they impart has been immortalized in the names of rocks and hills throughout Europe. In Scandinavia, the term 'gråstein' or greystone refers to grey

ABOVE **Scanning electron micrograph showing the contact between the crustose map lichen *Rhizocarpon geographicum* (dark grey) and granite rock (light grey) from Cumbria, UK. Long and sinuous fungal hyphae have penetrated several millimetres into the interior of the rock.**

lichens (species of *Aspicilia*, *Porpidia* and *Parmelia*) growing on granitic rocks. Perhaps the most striking contrasts in lichen communities are those occurring on limestones consisting of calcium carbonate compared with those on siliceous (quartz-rich) rocks. A wide variety of immersed lichens occur on limestones, often together with bright-orange species belonging to the genus *Caloplaca*, while siliceous rocks often support the yellow-green map lichen *Rhizocarpon geographicum*.

Metal-rich sites often lack conspicuous vegetation because most higher plants cannot cope with high concentrations of often toxic metals in the soil. A few plants are adapted to such sites (metallophytes) and form unique communities, which are tolerant of toxic metals such as copper, lead and zinc. These elements, incidentally, form the basis of several fungicides and algicides. So it is surprising that lichens also grow in these places. Bacteria also assist in the weathering of rocks rich in iron sulphides, e.g. 'fools gold', creating sulphuric acid. In these low-pH environments the rust-coloured lichen *Acarospora sinopica* and a range of other species are found. In contrast, the 'copper lecidea', *Lecidea inops*, is characteristic of more alkaline, copper-rich rocks.

ABOVE Lichens colonizing metal-sulphide-rich rocks, Litlfjellet, Gjersvik, Nord Trøndelag, Norway.

LEFT Close-up of crustose lichen community on slate, including *Caloplaca* cf *ignea* (orange) and *Acarospora* sp. (yellow). Sierra Nevada foothills, California, USA.

LEFT Slate boulder rock, Sierra Nevada foothills, California. The orange lichen is *Caloplaca* cf *ignea*; the yellow one an *Acarospora* species.

# Lichens in forests

## Deciduous woodlands

Ancient trees often support over 30 different lichens and the oceanic woodlands in the west of Scotland, UK, may contain more than 400 different kinds. Some lichens have limited powers of dispersal and are restricted to these old-growth forests so, generally, the older the woodland, the better it is for lichen diversity. Various lists of species that are characteristic of ancient woodlands have been compiled in different countries. These 'indicator' species can be used to assist in identifying forests of high conservation value.

Many factors influence lichens growing on trees including aspect, slope, bark texture and acidity, and competition from mosses. Throughout the life of a tree there are changes in its bark, its ability to store water, acidity (pH), nutrient status and light levels. Older trees usually provide a greater variety of microhabitats, allowing many different types of lichens to colonize. Sometimes two apparently very similar trees standing side by side may support different lichen assemblages, although this is unusual. When it does occur, it may reflect subtle differences in bark chemistry or pH. This is one of the thrilling aspects of searching for lichens in the field – there are sure to be surprises no matter how experienced you are. External factors also play an important part in influencing which lichens grow on bark. For instance, trees close to cement quarries often develop lichen communities that are typical of those

### Why do different lichens grow on trees?

A single tree provides diverse habitats for lichens during its life. Communities change through natural succession and due to changes in the environment of the tree. Some take up to 30 years to develop to maturity. Top left to right: *Lobaria pulmonaria* (ancient trees), *Xanthoria parietina* (yellow) and *Physcia adscendens* (grey) (nutrient enrichment), *Usnea inflata* (twigs and branches), *Ramalina fraxinea* (exposed trunks). Bottom left to right: *Leptogium cochleatum* (sheltered, mossy trunks), *Peltigera horizontalis* (mossy tree bases), *Lecanactis abietina* (crevices, dry bark), *Caloplaca luteoalba* (wound tracks). Locations of species on the tree may vary.

on calcareous rocks. This has been called the 'alkaline dust' effect. Air quality, nutrients from farming, light levels and air humidity are also very important influences.

## Northern coniferous forests

Northern conifers are excellent for lichens, the older the trees the better they are for lichen diversity. These forests are home to the famous and striking *Usnea longissima* *(p.58)*. Popularly known as long-beard lichen or Methuselah's beard, it is longest of all lichens, occasionally unravelling to lengths of more than 3 m (9.8ft). It has a virtually circumpolar distribution in the

boreal coniferous woodlands of Europe, Asia and North America. Norway is its last major stronghold in Europe, where it occurs twisted around branches of spruce, *Picea abies*. It is believed *Usnea* spp. were the basis for the tradition of decorating Christmas trees with tinsel.

Normally occurring on old trees, *Usnea longissima* depends on old-growth forests but may occasionally colonize young trees where conditions are particularly favourable. In the Pacific North West of America, an important stronghold for this and many other lichens, *Usnea longissima* extends south to the coastal redwood, *Sequoia sempervirens*, zone in Northern California. This lichen occurs

ABOVE **Joffre Lakes, Provincial Park, British Columbia, Canada. The grey, hoary coniferous trees are covered in beard lichens, *Usnea*, *Alectoria* and *Bryoria* spp.**

RIGHT ***Usnea longissima* hanging from trees in South-east Alaska.**

within the range of the northern spotted owl, *Strix occidentalis*, and, like this conservation flagship species, is protected virtually throughout its range.

## Southern rain forests

Lichens are among the most diverse plant groups in rain forests in the Southern Hemisphere. In the Australian state of Tasmania these forests contain more than 200 macrolichens and at least as many crustose lichens, outnumbering flowering plants by about four to one. The lichens of rain forests throughout New Zealand, Tasmania and Southern South America are very similar and we

ABOVE This remarkable crustose lichen, *Psoroma durietzii*, is very unusual because it develops sorediate cephalodia rather than soredia containing green algae.

LEFT *Pseudocyphellaria glabra*, one of the most common rain-forest lichens in Tasmania and Chile.

believe that they all originated in the ancient continent of Gondwanaland, which split into the southern continents of today some 65 million years ago.

The best places for lichens in these rain forests are open places where the light levels are relatively high. Lichens are particularly well developed at higher altitudes where trees are less dense, but even at lower altitudes they may smother the taller branches that protrude above the canopy. The large species of *Pseudocyphellaria*, *Psoroma*, *Sticta* and *Nephroma* grow rapidly. These lichens contain cyanobacteria and can fix atmospheric nitrogen, so they may contribute significant amounts of nutrients to rain forests. Studies in New Zealand suggest the amount may be as high as 1–10 kg (2.2– 22lb) of nitrogen per hectare compared with 1–2 kg (2.2–4.4 lb) from rainfall. Because lichens grow quickly in these lush forests and have a rapid turn-over, they may often be seen decomposing on the forest floor. Field trials have shown that as much as 25% of *Pseudocyphellaria glabra*, one of the most common rain-forest lichens in Tasmania and Chile, may decay in a 4.5-month-period. Other lichens, particularly *Sticta* spp., may decay even faster. Because lichens contain nitrogen, largely in the form of proteins and chitin with smaller amounts of amino and nucleic acids, lichen litter can be a valuable additional source of nitrogen. Some species smell like rotting fish when damp because they contain high levels of amines, especially trimethylamine.

*Psoroma durietzii* is a remarkable crustose lichen of southern rain forests. It is very unusual because it develops sorediate wart-like structures (cephalodia) rather than soredia containing green algae. The fungus is therefore dispersed in soredia containing species of the cyanobacterium genera *Nostoc*. The young thalli have a rather amorphous shape and do not resemble their parent. They need to capture compatible green algae to form the typical organized thallus. Interestingly, the earliest cyanolichen known, the fossil lichen *Winfrenatia reticulata*, also forms soralia-like structures (see p.46).

# Tropical rain forests

Tropical rain forests are well known as major sites for the biodiversity of flowering plants. They also provide habitats for lichens, which grow on the shaded buttresses of giant trees and up into the canopy 20–30 m (66–99 ft) above the ground. While lichens contribute relatively little to the overall biomass, they do contribute to diversity. In lowland forests, crustose lichens containing cyanobacteria of the genus *Trentepohlia* as photobiont predominate, being able to tolerate the shaded, moist environment. In the canopy, inaccessible areas are still poorly explored in many regions, but it is here that macrolichens and other crustose lichens containing green algae may occur.

In some places tropical forests have remained stable for many thousands of years and in these places lichen diversity may be very high in small areas. In the understorey of rain forests in Costa Rica up to 300 species of lichen occur on leaves (foliicolous lichens) at a single site. There, the laurel *Ocotea atirrensis* has been found to support

50–80 different lichen species on a single leaf. To put this into perspective only two or three lichens grow regularly on leaves in Britain, particularly on those of the box tree, *Buxus sempervirens*. The high growth rate of tropical plants constantly provides new leaf surfaces for colonization so that many foliicolous lichens reach sexual maturity, producing fruiting bodies or propagules, within a 6–12 month period. Foliicolous lichens usually occur widely on a range of perennial leaves of vascular plants. Most species probably use the leaf only as

something to grow on but it is possible that lichens growing beneath the cuticle may also derive nutrients from the plant. At higher altitudes in tropical regions, where the forest is immersed in cloud for most of the day, lichen biomass and diversity are even higher, with tangled growths of lichens, mosses and epiphytic orchids hanging from every branch and twig. One *Elaeocarpus* tree that had fallen in Papua New Guinea supported 173 different lichen species, probably the highest number of lichens ever recorded from a single tree.

LEFT **High-altitude cloud forest. Monteverde, Costa Rica.**

TOP RIGHT **Epiphytic vegetation is luxuriant: entangled growths of orchids, bromeliads, mosses and lichens. Volcan Cacao, Costa Rica.**

MIDDLE RIGHT **Tropical forests are the best habitats for lichens growing on leaves. Leaf colonized by *Byssoloma discordans*.**

BOTTOM RIGHT **_Byssoloma discordans_, Costa Rica.**

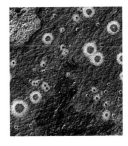

# Lichens in extreme environments

## Urban lichens

Generally we think of lichens as slow-growing organisms typical of pristine, natural environments but they grow in built-up areas too. A number of 'lichen weeds' occur in towns and cities throughout the world, often unnoticed by the human inhabitants who unwittingly crush them beneath their feet. Tarmacadam pavements, brick and concrete walls, roofs, trees, dustbin lids, old leather, rubber tyres and even rusting motor cars on municipal dumps may provide ideal habitats for lichens. Buildings may be extraordinarily rich for lichens, especially in clean air areas. Lichen ecologist Dr Oliver Gilbert discovered that 20% of lichens he recorded during a survey of urban waste sites in Britain were either new to the country, nationally rare or badly under-recorded.

Around 40 different lichens have been found growing on stained glass windows in churches dating back to the 15th and 16th centuries in Northern France. Surprisingly, lichens were also found growing on the insides of some windows where condensation could occur, but this is unusual. Some colours of glass are less susceptible to attack by lichens than others, particularly golden-yellow glass containing toxic silver salts. Churchyards are important oases for lichens. In lowland England, where natural outcrops of rock are absent, the churchyard is the most important site for lichens growing on stone. Some species rarely occur in other habitats. For example, in Britain the rare *Lecanactis hemisphaerica*, which is on the UK Red Data List of Threatened Lichens, is confined to plaster walls on ancient churches

FAR LEFT **Church colonized by grey lichens, Pembroke, Wales, UK. The more-or-less lichen-free zone is a result of run-off from the lead in the stained-glass windows but close examination reveals some lead-tolerant species.**

LEFT **Lichens growing on a calcareous gravestone. Pembrokeshire, Wales.**

LEFT **Lichens growing on an acid slate gravestone. Lundy Island, Devon, UK.**

in Southern England. In Britain as a whole, over 630 species (over a third of the total occurring in the UK) occur on or near churches, mainly on stone, but also on wood, trees and soil in churchyards, cemeteries and the surroundings of abbeys and cathedrals.

Metal-loving lichens may also grow on or under man-made structures. An interesting example is *Lecanora vinetorum*, which grows on vine supports in Austria, where it is sprayed routinely with the fungicide bordeaux mixture, which is used to control mildew on grape-vines. It is unknown elsewhere in the natural environment and may have evolved recently. Lichens may also be found beneath the drip line from barbed-wire fences, electricity pylons and crash barriers along motorways.

# Rocky coasts

Rocky coasts subjected to salt-laden spray and winds are extremely hostile environments for many plants, but lichens may excel there. If you live in temperate latitudes, next time you visit a rocky coast take a closer look at the rocks and marvel at one of the most dramatic examples of ecological zonation in a small area. Above the seaweeds and barnacles lie distinctive coloured bands of different lichens, each having a different tolerance of salt spray. Some lichens are regularly immersed by sea-water at high tides. Higher up the shore nutrient enrichment by birds is also important in determining what species of lichen can grow.

The lowest band is black and tar-like, colonized by the pyrenolichens, *Verrucaria* spp. Above that is a striking bright-orange

band dominated by *Caloplaca* spp. and *Xanthoria* spp. This finally gives away to a grey zone colonized by different crustose lichens and the shrub-like *Ramalina* spp. All bands may be telescoped within a few metres, depending on the steepness of the rocks, the extent of the tidal amplitude and the degree of exposure. The zonation is best developed in the Northern Hemisphere on steep south-facing coasts. On exposed islands, such as the Shetland Islands in northern Scotland, salt-laden storm winds allow colonization of the black *Verrucaria maura* to the summits of mountains *c.* 450 m (1480 ft) above sea level. Very occasionally, salt-laden winds blowing inland, particularly during south-westerly winter gales, may enable some coastal *Ramalina* species to colonize inland sites. A good example of this can be seen at the ancient prehistoric monument of Stonehenge, which lies some 50 km (30 mile) inland from Britain's southern coast.

As you travel closer to the Equator, a combination of intense heat and salt create conditions normally too hostile for even lichens to grow well at sea level and the characteristic colour zones disappear. Thus, on the Atlantic Islands of Madeira and Porto Santo, lichens grow only higher up the cliffs, at around 100–200 m (3300–6600ft) above sea level, and at about 300–400 m (9800–1300ft) on Ascension Island in the mid-Atlantic.

OPPOSITE **Lichen zonation showing black (oil-tar lichen, *Verrucaria maura*), and orange (*Caloplaca* sp. and *Xanthoria* sp.) bands. Pembrokeshire, UK.**

BELOW **The pyrenolichen, *Pyrenocollema halodytes*, immersed in barnacles. Note the numerous tiny black dots.**

## Oil

While British Petroleum was building the Sullom Voe Oil Terminal in the Shetland Islands, Scotland in the mid-1970s, a very minor local oil spill resulted in numerous reports of extensive slicks of oil washed up on the coasts all round the islands. When these reports were investigated it was found that, in all cases, the substance involved was in fact the black lichen, *Verrucaria maura*, sometimes called the tar-spot lichen. It seems that few people had noticed its existence prior to this accident. Oil spills can discolour lichens but the oil itself seems to be less toxic than the cleansing agents used in 'clean-up' operations. These may also disperse the oil more widely so that all surfaces are affected. Some recovery, at least of the commoner lichen species, usually occurs within 5–10 years of an oil spill.

BELOW Seasonal variation in different types of plant in reindeer rumen content at Hardangervidda, Norway.

BOTTOM A mule deer eating *Bryoria fuscescens* and *B. pseudofuscens* on a tree branch. Swan Valley, Montana, USA.

# Arctic tundra

Arctic tundra is dominated by lichens. Lichens are everywhere – on the ground, shrubs and rocks. The most conspicuous lichens are reindeer moss, *Cladonia rangiferina* and its relatives. In some areas lichen biomass may exceed 300 g/m³ (0.5 lb/yd³). Reindeer (caribou in North America), *Rangifer tarandus*, feed on a range of plants according to availability of palatable species, and reindeer moss and other lichens form a staple part of their diet, especially during winter months when lichens represent 60–70 per cent of their food intake. The animals reach the lichen by digging craters in deep, soft snow. Most organisms cannot digest the principal polysaccharides, lichenan and iso-lichenan, present in lichens but reindeer have a gut microfauna containing the necessary digestive enzymes. A single reindeer may graze up to 2160 m² (2580 yd²) during a 6-month winter period and it takes 10–20 years for a grazed area to regenerate once reindeer are removed.

## Radioactive reindeer

Lichens have been used successfully to monitor contamination by radioactive elements derived from nuclear bomb testing, the crashing of nuclear-powered satellites and accidents at power stations. When reactor 4 of Chernobyl's nuclear power station exploded on 26 April 1986, livestock herds throughout the Northern Hemisphere were threatened by radioactive contamination. Nowhere was the situation more acute than for the Norwegian Saami people, Laplanders who were formerly widespread in the arctic parts of Norway, Sweden, Finland and Russia. These nomadic people rely heavily on reindeer for their food, clothes and trade. Days after Chernobyl, lichens in Europe had accumulated levels of the radioisotope caesium 137 that were up to 165 times higher than levels recorded previously. The legal limit for sale of reindeer meat was originally 300 Becquerels (Bq)/kg (136 Bq/lb)

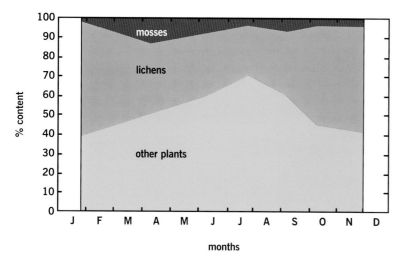

but this was subsequently raised in November 1986 to 6000 Bq/kg (2721 Bq/lb). Levels exceeding 10,000 Bq$^{137}$Cs/kg (4535 Bq$^{137}$Cs/lb) were measured in some reindeer. Even such dubious legislation made little difference to the Saami. In 1988 alone, 545 tonnes (536 tons) of reindeer carcass had to be disposed of as toxic waste. The consequences for the Saami have been severe. Their primary source of food is now completely contaminated and it is impossible for them to distinguish between reindeer that are massively over-contaminated and those passing government safety levels. As a result of the ionizing radiation, incidence of thyroid and other cancers in the local human population has risen. Trade in reindeer has been the monopoly of the Saami and the basis of their independent existence. Since the Chernobyl disaster more and more of the 19,000 remaining Norwegian Saami are turning to the government for support. Dependency on the state is resulting in the loss of generations of skills and cultural practices and is undermining the very existence of the Saami people.

• Lichens are very resistant to ionizing radiation. In experiments they survived 1000 rads per day for nearly 2 years from a distance of 8 m (26 ft) and continued to grow, although all vascular plants in the same experiment were killed, including the trees supporting the lichens. Man normally dies after a single exposure to 400 rads.

• There is great potential to use lichens as bioindicators for a range of radionuclides, not just caesium. Because they accumulate such high concentrations of these substances, they are easier to analyse than many other biological materials.

## Alpine peaks

Mountains are inhospitable environments subject to extreme temperature fluctuations, high UV levels and wind speeds, variable snow cover and short growing seasons. Lichens must cope with all these in order to survive. The high mountains of the Alps and Himalaya share broadly similar lichen communities to those of arctic regions, although some are highly localized and specific to particular regions. Lichens have been recorded at extreme elevations in the Himalaya. During a 1972 Yugoslavian–Himalaya expedition, *Xanthoria elegans* was recorded at 7000 m (23,000 ft) in the Karakorum and *Lecanora polytropa* was recorded on the south side of Makalu up to an altitude of 7400 m (24,300 ft). Interestingly, neither of these lichens is restricted to high altitudes and

BELOW *Umbilicaria virginis*, a 'rock tripe', from the Sierra Nevada, California, USA. The thalli of *Umbilicaria* spp. may grow to 15 cm (6 in) across.

ABOVE *Xanthoria elegans*
(bright orange) and other
lichens on exposed
boulders on the summits
of high mountains must
endure extreme climatic
conditions, including high
levels of UV-radiation.
Tirol Alps, Austria.

both occur frequently at much lower altitudes on silicate rocks throughout the world. These particular lichens are able to grow at these high altitudes almost certainly as a result of local nutrient enrichment by birds.

Snow beds, where a thick layer of snow offers protection from extreme cold for much of the year, are also interesting habitats for lichens. During the brief alpine summer snow beds become particularly wet, cold and dark environments. Lichens typical of these places are often associated with cyanobacteria, at least in their early stages of development, which they may swap later for a green alga. Thus *Solorina crocea* exists at first as tiny thalli tucked inside moss cushions and forms its typical thallus only when it later obtains a green alga. The ability of the blue-green photobiont to fix nitrogen is believed to be especially important in these challenging environments.

On the drier, exposed alpine heaths a straggling, chalky-white, worm-like lichen, *Thamnolia vermicularis*, occurs among mosses and small shrubs. This species is found in both the Northern and Southern Hemispheres. It is very unusual among macrolichens because its only known method of reproduction is by simple fragmentation.

# Antarctica

The Antarctic land area is vast, extending over about 14.5 million km$^2$ (5.6 million mile$^2$). However, only the maritime antarctic, representing less than 2% of the total area, is ice-free. Two rather inconspicuous flowering plants occur in the entire region and one is a grass. In the milder and wetter (*c.* 400 mm

ABOVE **Worm lichen,** ***Thamnolia vermicularis,*** **scrambling among mosses and alpine vegetation.**

RIGHT ***Solorina crocea*** **has a bright orange lower surface and immersed brown apothecia. Late snow patch, Alps.**

(16 in) precipitation/year) maritime antarctic, extensive shrubby lichen communities cover several hectares in favourable stony areas. About 200–300 species of lichens are thought to occur in the Antarctic as a whole, the dominant ones being around 25 shrub-like species belonging to the genera *Usnea* and *Bryoria*. Recent studies have shown that lichens may grow faster than we previously imagined, especially where there is some nutrient enrichment from birds, for example the giant petrel *Macronectes giganteus*. Lichens have also been found growing on fairly recent broken glass of soft-drink bottles. Life is fragile in this harsh environment but there are signs that lichens are adapting to global changes, such as ozone depletion in the outer layers of the atmosphere and the resulting increased amounts of UV-B radiation reaching Earth.

## Ozone hole

Dramatic loss of ozone in the outer layer of the atmosphere (stratosphere) was first noticed over Antarctica in the 1970s by the British Antarctic Survey in Cambridge, UK. The phenomenon is thought to be primarily the result of using chlorofluorocarbons (CFCs) as cooling agents in fridges, freezers and air conditioners. This depletion of the

BELOW **Shrubby lichens are dominant in coastal maritime antarctic areas.** *Usnea aurantiaco-atra* **heath. Livingston Island, South Shetland Islands, UK.**

ozone layer is serious because ozone absorbs some of the potentially harmful ultraviolet (UV-B) radiation from the sun. Although UV-B radiation contributes less than 1% of the total sunlight reaching the Earth, it is damaging to organisms, disrupting essential life processes of many forms of life. High levels also contribute to the development of skin cancer in humans and other animals. Despite the 1987 Montreal Protocol on Substances that Deplete the Ozone Layer and other agreements that aim to reduce global emissions of CFCs, the problem appears to be on a much larger scale than we anticipated. We all need to protect ourselves from excess levels of radiation that may cause skin cancer. Studying lichens and other living things that are naturally adapted to high levels of UV-B is therefore important.

## Natural sunscreens

Lichens have evolved ways of protecting themselves from UV-radiation. Levels of usnic acid in the upper cortex of several lichens, including *Usnea* (previously known as *Neuropogon*) *aurantiaco-atra*, in maritime antarctic lichen heaths vary according to the seasons and incident UV-B levels. This suggests that, like mosses, lichens may adapt to changing environmental conditions,

LEFT *Usnea* (*Neuropogon*) *aurantiaco-atra* in the herbarium of The Natural History Museum. One of 34 lichens collected by Charles Darwin during the voyage of the *Beagle*.

producing more usnic acid when levels of UV-B are high. Usnic acid levels in old herbarium samples are different to those in recent samples collected from the same areas, providing further evidence that lichens are a convenient tool for monitoring changes in UV-B radiation. Other pigments are present in the outer layers of lichens, which may also be involved in protecting the lichen from UV-radiation-damage. At least one pharmaceutical company is investigating these properties with a view to developing more effective sunscreens.

## Martian life forms?

If you were to travel inland in the Antarctic, you would notice that the further inland you go, the fewer and fewer lichens you find on rock exposures. Southwards along the Trans Antarctic Mountains there are about 28 species near 78°S; eight lichens at 84°S in a localized, very favourable site; and only two poorly developed lichens at 86°S (close to the South Pole). In such harsh conditions survival depends on whether lichens can photosynthesize at extremely low temperatures. If they cannot absorb enough moisture from the atmosphere under snow and cannot photosynthesize while frozen, then they are unlikely to occur. Lichenized fungi have evolved ingenious ways of coping with these harsh conditions.

RIGHT **Antarctic cryptoendolith.**

Among the most extreme environments on Earth are the intensely cold, interior antarctic valleys. Dry winds descend from the antarctic plateau, creating true desert conditions where maximum air temperatures never rise significantly above 0°C (32°F) and drop to near –60°C (–76°F). Extensive areas of rock and soil are without snow and ice cover, and daily temperature ranges may be extreme. In sunny periods with an air temperature of –40°C (–40°F) surfaces may reach 30°C (86°F). Mean annual relative humidity and precipitation are low. In this frigid desert there are no visible signs of plant and animal life. Yet inside translucent, porous rocks a narrow zone just below the surface provides a favourable microclimate for minute forms of life. Lichens known as 'cryptoendoliths' (organisms hidden within rocks) are dominant in this zone. They survive by penetrating between colourless crystal grains of granite and marble. A colourless upper layer of quartz grains protects them within the rock, and fungal hyphae are loosely associated with green algae in the black pigmented and inner white layers sandwiched between coloured layers. Common green algae of the genus *Trebouxia* have been identified in the green layer. Bacteria, including cyanobacteria, may also be present in the community but are not associated with fungal hyphae. There is some variation in colour sequences, the black pigmented layer sometimes occurring on top. This variation is thought to depend on environmental factors. Such lichens are usually reduced to a simple filamentous growth habit but they may, under favourable microclimatic conditions, produce small crustose thalli with highly organized, layered thalli bearing apothecia. Several genera, including *Acarospora*, *Buellia* and *Lecidea*, have been recognized. These lichens occur together with a range of other non-lichenized fungi, algae, cyanobacteria and other bacteria.

Survival of antarctic cryptoendoliths depends on a precarious balance between biological and geological factors. Any unfavourable shift in conditions may result in death and formation of microfossils – tell-tale signs of the existence of previous life. Indeed it has been suggested that these Antarctic environments are the closest terrestrial equivalents to Mars, leading to speculation that lichens might be found on Mars itself. Several research groups are busy studying these strange life forms, and the associated minerals and weathering patterns, and comparing them with rocks from Mars.

# Hot deserts

Arid desert regions, where flowering plants are absent, may support lichens in abundance because lichens can absorb enough water from damp air to become active physiologically. Thus in Australia, *Ramalina maciformis* can photosynthesize at a relative humidity of 80%. Such humidities are frequent in arid areas because the temperature falls rapidly at night causing dew to form. The thallus uses the dew to hydrate overnight and at dawn increasing light causes a rapid rise in photosynthesis. The thallus dries out as the temperature rises and photosynthesis ceases (see p. 25). Coastal deserts provide ideal conditions for lichens

ABOVE *Teloschistes* fields
in the Namib Desert,
Namibia.

because the sea moderates temperature extremes and often creates high humidity, fog or dew. Shrubby lichens, belonging to the genera *Roccella*, *Ramalina* and *Teloschistes* are most frequent, often dominating areas where few or no flowering plants are found.

Although the Namib Desert is the most arid region in Southern Africa, a narrow coastal strip 1000 km (620 mile) long is bathed in coastal fog for up to 285 days each year. Even when there is no fog, relative humidities may remain at 100% for most of the day. This is due to the upwelling of cold nutrient-rich antarctic waters meeting the warm Benguela current along the coast. The coastal gravel plains of the central Namib Desert are covered by lichen fields dominated by the bright-orange, shrub-like *Teloschistes capensis*, which can grow to 10 cm (4 in) tall. Research by a German team of scientists, led by Professor Otto L. Lange, at the University of Würzberg, measured photosynthesis in the field and demonstrated that lichen growth is restricted to sunrise on foggy days, when water availability, light intensity and temperature are optimal. But the lichens do not occur throughout the entire fog zone, suggesting that other factors might be important. Close by the lichen fields 100,000 Cape fur seals, *Arctocephalus pusillus*, bathe in these productive coastal waters. An intriguing possible explanation currently being explored by Dr Peter Crittenden at the University of Nottingham, UK, is that nitrogen produced in excreta from seals is also important for lichen growth in this habitat.

The Atacama Desert along the coasts of Peru and northern Chile is perhaps the driest region in the world in terms of measurable precipitation. Here the northward-flowing Humboldt current meets deep ocean waters off the Atacama coast, producing a mild climate. Beard-like lichens flourish and these efficient absorbers of moisture from fog and dew may even help the cacti around them to obtain water.

Several strange lichens occur in arid and semi-arid areas, including unattached lichens or 'Wanderflechten' as they are called in Germany. In 1829, during the war between Russia and Persia, a town on the Caspian, was suddenly covered by a lichen, which literally fell from heaven. It was made into bread and eaten by the starving people. Perhaps this lichen, *Lecanora esculenta*, was the manna of the Israelites during their flight from Egypt.

LEFT **Lichens growing on cacti in the Atacama Desert, Chile.**

# Biomonitoring

Using living organisms to assess environmental quality dates back a long time. A well known early example was coal miners taking canaries down the pits; if the canaries became drowsy then it was sign that the odourless gas methane was reaching dangerous concentrations. Monitoring the pollution status or health of the environment using lichens has been carried out extensively for over 30 years and a large literature has been published numbering several thousand articles.

• Bioindication uses an organism to obtain information on the quality of its environment. The organisms used are called bioindicators and they serve to detect and identify the effects of pollutants and other forms of environmental disturbance as an alternative to direct measurement of the pollutant or disturbance.

• Biomonitoring is observing an area over a

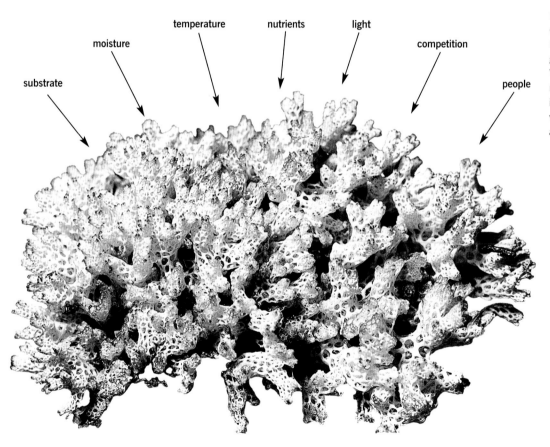

substrate   moisture   temperature   nutrients   light   competition   people

LEFT **Lichens, like other organisms, depend on a number of factors for growth and survival. When designing any monitoring programme it is important to consider which other factors may also influence lichens.**

period of time with the help of bioindicators. In other words, the difference between bioindication and biomonitoring is the same as that between taking a photograph and making a movie film.

Lichens are useful bioindicators because they are widely distributed on a global scale, they form perennial bodies, are usually long-lived and they have a habit of concentrating elements from their environment. They may be used as indicators in several different ways. These include assessing physiological, biochemical or morphological changes, or changes in community structure through extinction or species substitution. The dual nature of the lichen association and its sensitivity to environmental disturbance is a major reason why lichens are useful. If the delicate balance between symbionts is upset, this can lead ultimately to the death of the lichen. Not all lichens respond in an identical fashion; different species show varying levels of sensitivity to particular environmental factors.

Air quality (concentrations of sulphur dioxide, fluoride, ammonia etc), metal contamination, the conservation status of woodlands and the ozone hole have all been monitored using lichens. While lichens do not replace monitoring techniques using instrumental recording gauges, they do permit a higher sampling density and are rapidly gaining favour throughout Europe as a low-cost option to supplement data obtained from instrumental recording. Indeed, air-pollution monitoring using lichens is becoming mandatory in some European countries (Germany, Italy and Switzerland).

# Smogs

The smogs that were, and still are in places, such a feature of the industrialized cities around the world, have had a serious impact on lichen diversity. Formed as the result of burning of coal and fuel oil, and also by vehicle emissions, the sulphur dioxide ($SO_2$) component is the most harmful to lichens. So dramatic was the decline of lichens that the term 'lichen desert' was coined to describe areas in the centres of towns where trees are devoid of leaf- and beard-like lichens. In the late 1960s and early 1970s semi-quantitative, biological scales were devised to estimate $SO_2$ levels in the atmosphere. Lichen biodiversity

BELOW **Map of air quality based on 'Our Mucky Air' project produced by school children in England and Wales, who mapped lichens around their homes using a simple scale.**

**Mean winter $SO_2$ ($\mu$g/m$^3$)**

ca. 70–170

ca. 40–70

ca. 35–40

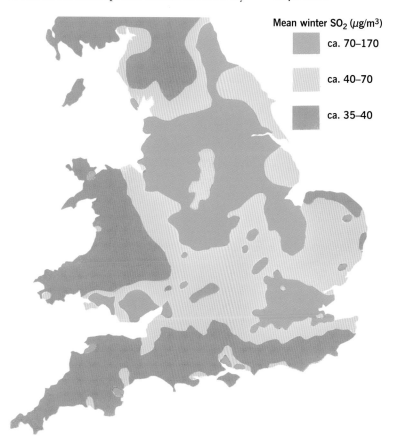

indices were calculated, which were found to correlate with pollution levels and human health throughout Europe, Japan and North America. Analyses of the sulphur contents of lichens and the relative abundance of different sulphur isotopes compared with substrate and source emissions have confirmed that lichens absorb a much higher proportion of sulphur from the air than do higher plants. Involvement has been at many different levels, from local to national,

BELOW **Historical sample of** *Usnea articulata* **collected from Enfield Chase, London towards the end of the 18th century.**

BOTTOM *Lecanora conizaeoides*, **the pollution lichen. Formerly the most abundant lichen in many industrialized areas of lowland Europe, it is now declining.**

involving a range of user groups from school children to scientific researchers.

Excessive levels of pollution will result in the ultimate loss of a lichen species, or change in species composition of a community, but this is normally preceded by various physiological reactions in the individual lichens themselves. Sulphur dioxide dissolves in rain-water, or in moisture within lichens, producing a variety of sulphur compounds. These are accumulated within the lichen and may have an effect on lichen metabolism, especially that of the photobiont, disrupting nitrogen fixation, respiration and photosynthesis. Physical manifestations include a bleaching of thalli (following loss of chlorophyll in algal cells), development of a red colouration (as a result of degradation of lichen substances) or a blackening, stunted growth. The lichen thallus may also develop lobules, decline in growth rate and fail to produce fruiting bodies.

National and local herbaria have provided valuable clues on the past distribution of lichens. Species containing cyanobacterial photobionts, for example *Lobaria* spp. and associated genera (*Sticta* and *Pseudocyphellaria*) are especially sensitive to sulphur dioxide. By contrast, the pollution lichen *Lecanora conizaeoides* is extremely tolerant and does not occur in non-polluted regions.

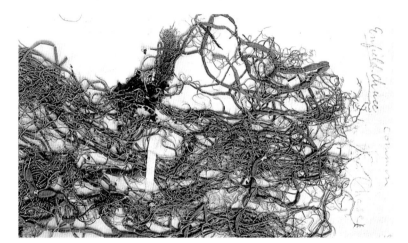

## Clean Air regulations and lichens

The UK Clean Air Act of 1956 and energy policy changes resulting in the use of cleaner fuels to achieve limits set by EEC (European Economic Community, now European Union) directives, have dramatically reduced national emissions of sulphur dioxide – by 80% since 1962. Today lichens are returning to many inner city areas following these dramatic decreases, a trend mirrored elsewhere in many parts of the world. The famous Jardin du Luxembourg in Paris, where in 1866 William Nylander made his first pioneering observations on the decline of lichens, today supports a good range of species. Seventy-two lichens were recorded recently from Kew Gardens, London – at the height of pollution fewer than six were found.

Lichen recovery has been slow because of a combination of factors. These include distance from remaining colonizing sources and acidification persisting on the bark of older trees. Where air pollution levels have decreased, recolonization by lichen species does not follow in the reverse order of losses due to pollution. Some pollution-sensitive species colonize more rapidly than more pollution-tolerant ones. Lichens that are able to colonize by asexual processes (soredia and isidia) return particularly quickly. The distribution range of *Usnea* spp. contracted to about two-thirds of its former extent in Britain at the height of the industrial revolution (see right). In 1970 *Usnea* started to reinvade the areas where it had been lost, a process that is still ongoing. In the Netherlands, following decreases in sulphur dioxide levels, species tolerant of nitrogen-

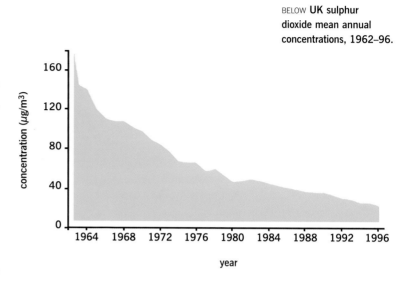

BELOW **UK sulphur dioxide mean annual concentrations, 1962–96.**

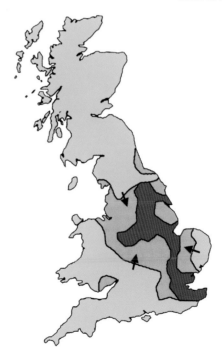

BELLOW **Distribution of *Usnea* in the UK.**

always present

reinvaded 1970-1998

disappeared 1800-1900, still absent

enriched environments caused by ammonia pollution (nitrophilous species eg. *Xanthorion* communities) have recolonized rapidly. It appears that throughout Europe, as sulphur dioxide levels decline, other pollutants may start limiting lichen recolonization. Incredibly, the pollution lichen, *Lecanora conizaeoides*, formerly ubiquitous in many industrial areas in Lowland Europe is becoming rare in places where it used to be the only lichen present.

Acid rain, resulting from the dissolving of sulphur dioxide and oxides of nitrogen, may cause increased bark acidification. Lichens that are tolerant of acidic conditions, such as *Pseudevernia furfuracea* and *Bryoria* spp., are favoured. Species containing cyanobacterial photobionts, such as *Lobaria* spp., are among the most sensitive to acid rain and are either disappearing from polluted sites or else becoming confined to trees with a high bark pH, such as ash, *Fraxinus excelsior*, which may help to neutralize acidification.

BELOW **Shrubby *Ramalina farinacea* and other lichens belonging to the colourful *Xanthorion* community are invading many areas in Britain following improvements in air quality. Burnham Beeches, West London.**

# Polluting air particles

Tiny particles suspended in air are among the most harmful of all pollutants for human health today. Particles less than 10 μm (0.01 mm), the so-called PM10s, are the most serious problem because they so readily enter the alveoli in our lungs. Metal-rich emissions, mostly insoluble particulates from industrial processes and motor vehicles, may become trapped on or in a lichen thallus. Lead, zinc, cadmium, nickel, copper, mercury and chromium, which are toxic to many living organisms, may accumulate to high concentrations in lichens. The lichens often appear to be unharmed but different lichen species vary in their sensitivity to metals and their ability to accumulate them. The amount of metal a lichen species accumulates also depends on the concentration and availability of metals in the environment. Biomonitoring studies designed to study particulate metal accumulation have used macrolichens most often, particularly shrubby ones with their high surface-area : volume ratio. They are also easy to sample and analyse. Lichens growing on trees are more useful than those on rocks because of potential contamination from a mineral substrate. Lichens are valuable in identifying the source of particulate emissions, locally as well as for whole regions or countries.

By analysing levels of metals in lichens we can identify where metals have come from and how far they are dispersed, especially where metals are associated with a particular industrial process. For instance, the metals vanadium and nickel may indicate petroleum combustion, for example from power stations burning fossil fuels, and lead may indicate exhaust fumes produced by cars running on leaded petrol. Because of their ability to accumulate high concentrations, lichens are excellent monitors of metal contamination, offering certain advantages to conventional air sampling. Several analytical techniques are available including inductively coupled plasma emission spectrometry (ICPMS) and electron probe micro analysis (EPMA), methods requiring only minute amounts of material.

RIGHT **Vanadium and lead contents in oak-moss,** *Evernia prunastri*, **at increasing distance from industrial plants in Pembroke, Wales, UK. The anomalous higher lead value at 22 km is the result of sampling close to a roadside.**

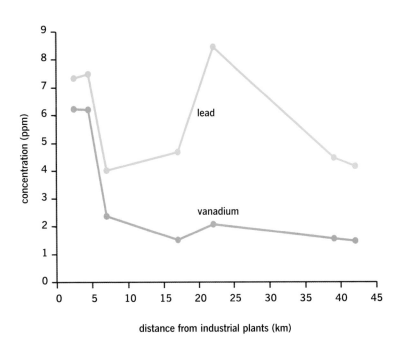

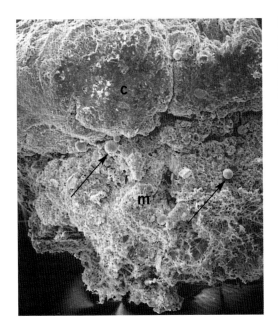

LEFT **SEM of** *Acarospora smaragdula* **(left), showing metal particles on cortex (c) and within the medulla (m). The extent of chemical contamination is most clearly seen from the bright areas of the right image. The smallest particles (PM10s) contain lead and are particularly damaging to human health. Arrows show same particles. Sampled near to ore-processing plant, Zlatna, Romania.**

# Remote sensing

Arctic systems are extremely fragile and vulnerable to disturbance and pollution. The properties of lichens offer opportunities to monitor these problems using remote sensing techniques, for example by using Landsat satellite images. Reindeer moss and other lichens containing the substance usnic acid are dominant in the vast arctic tundra. Usnic acid absorbs and transmits light differently compared with plants that lack this substance, so it is possible to monitor

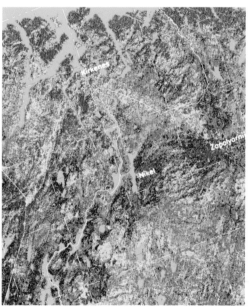

LEFT **Landsat maps of vegetation cover over the Kirkenes (North Norway) and Pechanga (Russia) area in (left) 1973 and (right) 1992. The colour differences are due to change from lichen heath/forest (yellow/light green) to bilberry (*Vaccinium* spp.) and meadow forests (dark green) and bare rock, eroded heath and contaminated (damaged) areas (purple).**

changes in the vast arctic and alpine lichen heaths from *c.* 35,000 km (21,700 mile) above the Earth's surface. Remote sensing can detect a lichen cover as low as 30%.

One example of the power of this method was the discovery that increased production from nickel processing from industries in Nikel and Zapolyarnijon on the Kola peninsula in North-west Russia had caused

severe damage to vegetation in Southern Varanger (Kirkenes) in Norway and the Pechenga municipality in Russia. Lichens were the most seriously affected and many lichen heaths have been replaced by bilberry, *Vaccinium myrtillus*, or by bare ground. Landsat data confirm that an area of 5000 km$^2$ (1930 mile$^2$) was damaged by air pollution in 1988 compared with 400 km$^2$ (150 mile$^2$) in 1973, during which time output of sulphur dioxide increased from 207,000 to 406,000 tonnes (204,000 to

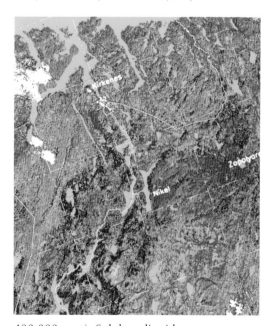

400,000 tons). Sulphur dioxide output was reduced to about 233,000 tonnes (230,000 tons) in 1993.

One of the main changes in the vegetation cover maps produced for the different years was that the area with lichen heath and lichen forest vegetation decreased from *c.* 37% of the area in 1973 to 10% of the area in 1992 with subsequently an increase to 15% in 1994. This increase may be due to reduced emissions. The emissions from the Nikel and Zapolyarnij chimneys are huge for such a small area, equivalent to roughly 5–10 times the total amount of sulphur dioxide released for the whole of Norway. Elsewhere, reindeer moss heaths in southern Scandinavia are contracting in their range as a result of changing farming and herding practices of reindeer, and possibly also due to increased levels of nitrogen compounds.

# Forest health

## Disappearing *Lobaria*

The lichen commonly known as tree lungwort, *Lobaria pulmonaria*, is widely distributed, occurring in Africa, Asia, Europe and North America. In most of its range it depends on undisturbed, pristine ancient forests that have never been felled completely. This attractive species is often used as a flagship species to help conserve primeval forests because it is readily recognized by

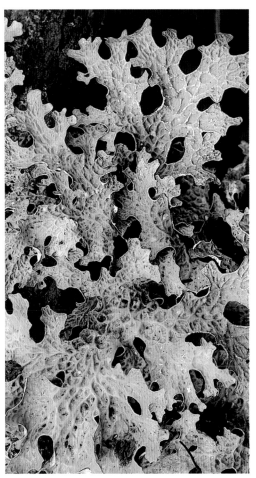

FAR LEFT *Lobaria pulmonaria* on sessile oak, *Quercus petraea*, Scotland, UK.

LEFT *Lobaria pulmonaria*, commonly known as the 'tree lungwort' because it resembles the lobes of a lung, Scotland, UK.

naturalists and foresters. During the last century this species disappeared from many parts of its range. This was partly because of changes in forest management, replacing forests of native species containing scattered veteran trees with even-aged plantations, often of non-native trees.

*Lobaria* species are also extremely sensitive to acidification, either from atmospheric pollution deposited as acid rain or to loss of ancient trees with base-rich bark. Dr Oliver Gilbert of Sheffield University, UK, found that under the influence of acid rain, *Lobaria* is unable to thrive on trees with an acid bark in parts of Northern England and has become restricted to trees, such as ash *Fraxinus excelsior*, which have a bark with a higher pH and buffering capacity. A national project set up across Britain by Peter James to record changes in species of *Lobaria* between 1986 and 1990, identified long-range atmospheric pollution as one factor influencing growth and also identified recovery in areas where atmospheric conditions had improved. Dr Christoph Scheidegger and his team are currently researching the decline in Switzerland, where, particularly on the Central Plateau, this lichen is very rare and may become extinct unless efforts are taken to conserve it. Scheidegger has pioneered a way of transplanting small lichen fragments (soredia) of *Lobaria pulmonaria*, wedging them between two small pieces of surgical gauze and stapling them to mossy tree bark. He has found, to his surprise, that it takes 12 months for the first lobules to appear, at least in Switzerland. Another problem for *Lobaria*

is that some of the very small remaining populations have very little genetic diversity, which means that they are even more vulnerable to extinction.

*Lobaria pulmonaria* is not threatened everywhere. Elsewhere in the clean air of the idyllic fairy-tale islands of the mid-Atlantic, which are often shrouded in mist, it is virtually a weed, occurring even on introduced trees and bramble stems.

## North American forests

Scientists and regulators like precise measurements when they have to evaluate the impact of disturbances on ecosystems. The instruments developed to analyse and take precise measurements of pollutants are expensive, however, and their cost can be prohibitive in large-scale projects. Biomonitoring is a convenient, relatively low-cost method to assess the effects of pollutants over large areas. The first large-scale effort in the world is the Air Quality Biomonitoring Program, which was established in the US National Forests of North-west Oregon and South-west Washington to meet federal and agency responsibilities to detect and describe air pollution impacts and protect forests. The programme is improving our understanding of the dynamics of lichen ecology, the use of lichens as indicators of forest health, stability and biodiversity.

Specially trained field personnel are used to follow the national Environmental Monitoring and Assessment Program/Forest Health Monitoring (EMAP/FHM) protocol developed for lichens by Dr Bruce McCune at Oregon State University in 1993. Plots are

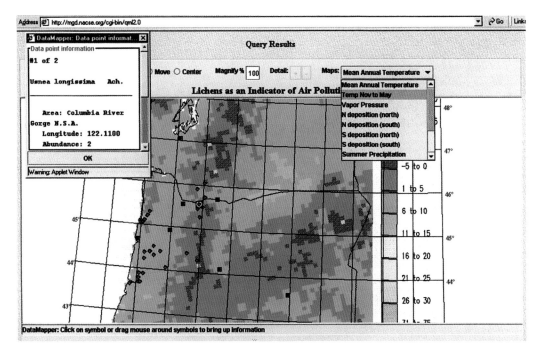

LEFT **On-screen query for**
*Usnea longissima* **using
the USDA Forest PNW
Lichens and Air Quality
on-line database.**

normally monitored on a 9.6 km² (3.7 mile²) grid on a 4-year basis. The lichen data, along with data on rainfall, temperature, pollution concentrations and other factors, are fed into a database linked to a geographical information system (GIS). Field crews regularly collect lichens and mosses for tissue analysis at all plots they visit (one-quarter of the total each summer). Concentrations of 27 elements are measured for *c.* 10 species region-wide. Field teams also carry out a survey of macrolichens on trees at each plot, including estimating how abundant they are. A programme co-ordinator maintains quality control. Work in all the forests uses the same guidelines so it is possible to compare results statistically in different forests, and generate reports and maps summarizing current air quality and how it is changing over time, and

changes in forest ecology. It is particularly important to be able to distinguish natural changes from detrimental effects caused by pollutants and other forms of disturbance.

Anyone can query the national forest lichen database for the US Pacific Northwest from the world wide web. You can search on different criteria. For instance, if you search on 'substrate' and type in PICO or PIPO (for *Pinus contorta* or *Pinus ponderosa)*, you can find out where the species occurs on that particular substrate. You can also search for a particular plant association (vascular plant community), for individual lichen species, or entire lichen communities. You can also plot individual lichen distributions on base maps showing precipitation or temperature data. This is a powerful tool with great potential to be developed in other parts of the world.

## The *Erioderma* story

In Northern latitudes south of the tundra regions, coastal coniferous forests support their own characteristic lichen vegetation. In Norway, the widespread Norway spruce, *Picea abies*, forms the climax vegetation and in Newfoundland, Canada, the black spruce, *Picea mariana*, takes its place. In certain cool, highly humid parts of the boreal forests on both sides of the Atlantic ocean, *Picea* is home to a rare lichen, *Erioderma pedicellatum*, one of the most vulnerable known to human disturbance. In central Norway, the lichen has declined and is now known only from a single tree in a damp spruce woodland. Its demise is an example of

BELOW ***Erioderma pedicellatum*. A very rare lichen in Norway currently known from a single tree in Central Norway.**

lessons to be learned on how legal protection may not guarantee the success of a species unless its biology and ecological requirements are understood. In Sweden the lichen was protected by law and remained only in one protected area in Northern Värmland. Unfortunately, adjacent woodland was felled, the lichen disappeared and is now extinct in Sweden. It disappeared probably because the felling of the woodland resulted in a lowering of the water table and subsequent decline in humidity of the area.

Fortunately, the situation in Newfoundland, home to the largest populations of *Erioderma pedicellatum* in the world, is not so bleak. Around 4800 individuals remain there, with

300 at one site. Newfoundland has a long history of forestry but certain areas remain largely intact. In 1996 the International Association of Lichenology (IAL) and International Union for the Conservation of Nature (IUCN, now IUCN–The World Conservation Union) alerted local authorities and, ultimately, the Premier of Newfoundland, to the importance of the Lockyers region for this rare lichen and succeeded in diverting a logging track from damaging populations. About $CAN400,000 were raised to help safeguard the species. Local resident and scientist, David Yetman, hired as a project biologist, is leading a team of local volunteers employed through Youth Services Canada to monitor populations.

LEFT *Erioderma wrightii*, a related species. Submontane rain-forest, Chiriqui, Panama. The wet thallus is swollen and covered in a felt of tiny hairs visible on the margins of the fruiting bodies.

ABOVE **Volunteer monitoring** *Erioderma pedicellatum* in Newfoundland, Canada.

# Prospecting and dating

## Metal prospecting

It has been known since Roman times that areas rich in minerals support characteristic floras. Initially, prospecting for ore minerals was based on searching for particular indicator plants or looking for differences in plant communities and their cover. Later, biogeochemical methods of exploration were developed, which rely on chemical analysis of vegetation. Exploration geologist Steve Czehura was among the first to recognize the practical value of lichens as biogeochemical indicators. In the Lights Creek District in the Plumas Copper Belt of northern California, USA, he identified unusually high levels of copper in the bedrock based on the colour of *Lecanora cascadensis*. Its colour varied from light (*c.* 1% Cu in ash) to dark malachite green (> 4% Cu in ash), corresponding to a copper content in the bedrock of 50–1000 ppm and > 2000 ppm, respectively.

Metal-loving lichen communities provide information on the types and concentrations of metals present in ore fields and also testify to our industrial past, where today spoil heaps cover the ground. Clues to the existence of ore deposits have remained buried in herbaria throughout the world for centuries. The significance of the green colour of *Lecidea theiodes*, which was first described

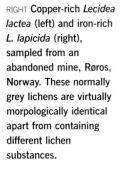

RIGHT **Copper-rich** *Lecidea lactea* **(left) and iron-rich** *L. lapicida* **(right), sampled from an abandoned mine, Røros, Norway. These normally grey lichens are virtually morpologically identical apart from containing different lichen substances.**

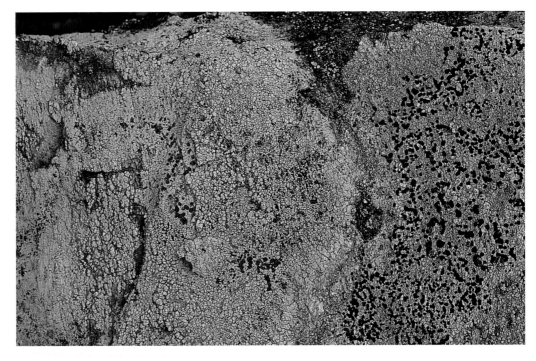

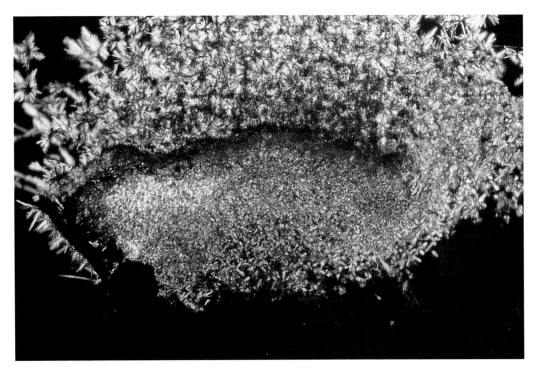

LEFT Sections of the green form of *Lecidea lactea* seen under the microscope in potassium hydroxide solution. Red, needle-shaped crystals show that norstictic acid is present in its upper copper-rich cortex.

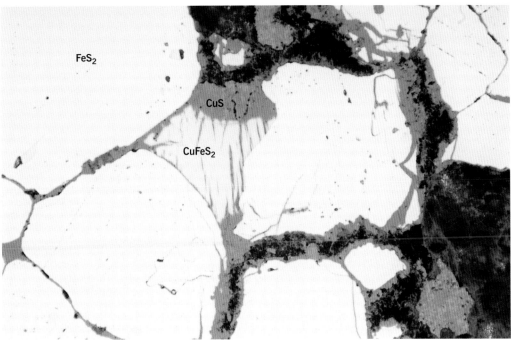

LEFT Rock section showing ore minerals. Copper ore, chalcopyrite ($CuFeS_2$) weathers forming covellite (CuS), releasing copper and iron; iron pyrite ($FeS_2$) releases iron. Lichens may accumulate copper and iron released by weathering.

by the Norwegian Christian Sommerfelt in 1826, from a copper mine in Saltdalen, Northern Norway, remained unknown for 150 years. It is now recognized as being an environmental modification of the widespread grey *Lecidea lactea* growing on copper-rich rocks. This was discovered by chance when collections of *L. theiodes* from Norway were examined in a scanning electron microscope (SEM) using an electron probe to analyse which elements were present. Surprisingly, the upper layer was found to contain copper, suggesting its colour was due to the copper rather than a normal lichen pigment. How is copper fixed in the lichen? By irrigating sections of the lichen with potassium hydroxide under a microscope, the characteristic red needles formed by norstictic acid appeared in the identical layer where the copper was detected. This suggested that copper might react with norstictic acid. Leading lichen chemist, Professor Jack Elix, at the University of Canberra, Australia tested this hypothesis. Analyses of other lichens confirmed that the type of acid present in the lichen determines whether a lichen may form metal complexes or not. This can be seen in the field, e.g. where green forms of *Lecidea lactea* ('*L. theiodes*') are growing next to *L. lapicida* (see p.88). These two lichens are virtually identical apart from containing different substances. Norstictic acid, present in *Lecidea lactea* and stictic acid, present in *L. lapicida* have identical chemical structures apart from a single functional group. This group determines whether or not they can form a complex.

In Europe about 12 crust-like lichens are now known that turn greenish yellow when growing on copper-rich rocks. These rocks normally contain copper ore, chalcopyrite ($CuFeS_2$), which weathers releasing copper and iron. Lichens growing nearby can then accumulate copper and iron. Lichens containing norstictic acid have been found to accumulate copper up to c. 5% dry weight.

Copper accumulation is therefore not simply due to mineral fragments being incorporated in the lichen thallus but depends on the kind of lichen species and the substances they contain. The research shows that metals can dramatically alter a lichen's appearance, so much so that it may be described as new species. Search for herbarium samples and, who knows, you may discover an ore deposit!

# Dating surfaces

Lichens grow very slowly compared with most plants. Although growth occurs in three dimensions, it is normally expressed either as an increase in radius (leaf and crust lichens) or an increase in tip length (shrub and beard lichens) – in both cases normally as millimetres per year. For instance, the crust-like *Rhizocarpon geographicum* grows around 0.5 mm/year (0.2 in/year) and the leaf-like *Parmelia caperata* 5 mm/year (0.2 in/year). The Californian *Ramalina menziesii* may attain rates of 9 cm/year (3.5 in/year), but such high growth rates are rare.

In many cases, the most active growth is restricted to the margins of the lichen thallus. By measuring the radial growth of circular lichen thalli you can work out both how old a lichen is and the probable age of the colonized surface. Dating surfaces using lichens (lichenometry) has often been used to date the retreat of glaciers, most often using crustose lichens on open rock surfaces. *Rhizocarpon geographicum*, abundant in such habitats, is the lichen studied most frequently. Its growth rate varies with age, young thalli growing faster than medium thalli and old large thalli growing even more slowly.

Obviously, dating any surface is limited to the potential age of the surface, which may be decades or perhaps centuries. It has been calculated that a few alpine lichens may live to over 1000 years, perhaps as much as 4500 years, rivaling the oldest flowering plants. Because lichen growth may be influenced by several factors it is important to design the study carefully, for example by selecting a single species growing in similar habitats and choosing several individuals to ensure that any variation in the data is statistically significant. Lichenometry is most useful where results can be checked against surfaces of known age. Churchyards are ideal control sites where they are present, because many gravestones are dated.

Dating has been used in:
• Dating moraines and the retreat of glaciers.
• How frequently earthquakes occur in Central Asia.
• Ancient monuments: e.g. the famous stone sculptures on Easter Island were found to be *c*. 400 years old.

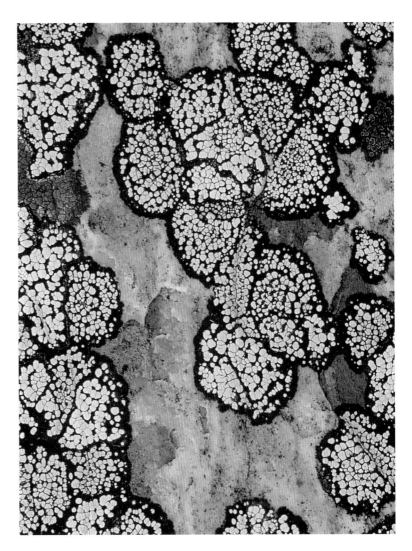

ABOVE **Map lichen,** *Rhizocarpon geographicum* **on rock, Glacier National Park, USA. A familiar lichen on silicate rocks, it consists of minute yellow-green islands, which contain the pigment rhizocarpic acid, growing on a black layer lacking algae.**

# Economic uses

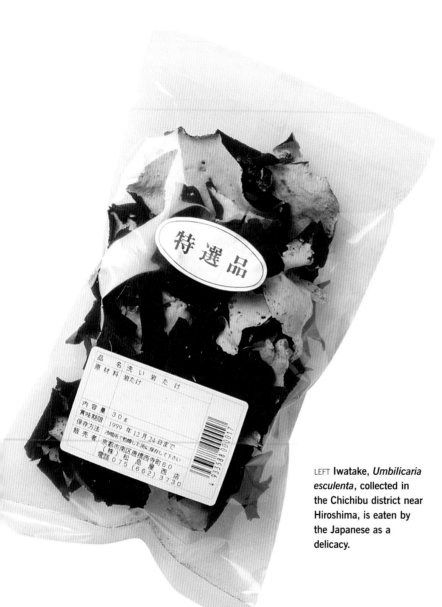

品

特選品

品名　洗い　岩たけ

原材料　岩たけ

内容量　30g

賞味期限　1999 年 12 月 24 日まで

保存方法　冷暗所で乾燥した所に保存して下さい

販売者　　京都市南区唐橋西寺町60

　　　　　（株）松島屋西店

　　　　　電話075（662）3730

LEFT **Iwatake, *Umbilicaria esculenta*, collected in the Chichibu district near Hiroshima, is eaten by the Japanese as a delicacy.**

Lichens are steeped in folklore and have been used in medicine from the time of the early Chinese and Egyptian civilizations. Humans have also used lichens for many purposes, from food to clothing. The principal limitation in developing the economic potential of lichens is their slow growth rate in culture. Large-scale collection in the field is nowadays unacceptable in many parts of the world because of the need to conserve natural resources. The greatest amounts of lichen used commercially appear to be in the perfume industry, amounting to many thousand tonnes annually, but it is hard to obtain reliable current data. Some 2000 to 3000 tonnes (1970 to 2950 tons) of the reindeer moss lichen *Cladonia* subgenus *Cladina* is collected every year for the construction of Christmas and graveyard wreaths, and for use in model building.

Industry has long relied on microorganisms as a source of pharmaceuticals, industrial chemicals and bioremediation procedures. Screening microbes isolated from the environment for the production of biologically active compounds, and controlled fermentation to manufacture large quantities of active molecules, have built a massive industry approaching $US200 billion per year in sales. However, this is based on exploration of less than 1% of microbes in the environment. The majority of microorganisms have been difficult or impossible to cultivate under artificial, laboratory conditions. An enormous wealth of bioactive compounds remains to be discovered and put to use as pharmaceuticals. Although lichens have been used by humans for centuries, no compounds are currently used on a commercial scale to rival the drugs aspirin and taxol, which were derived originally from willows, *Salix* spp., and the Pacific yew, *Taxus brevifolia*, respectively. In view of their high chemical diversity and given recent advances in genetic engineering, potential clearly exists for the commercial exploitation of lichens, testified by the current research activity of groups in Canada,

Japan and the UK searching for novel pharmaceuticals and agrochemicals.

# Dyes

Of all lichen dyes used by humans, none has attained greater historical and commercial significance than those of the coastal lichens belonging to the family Roccellaceae, commonly called orchella weed or orchil. They are mentioned in the Old Testament and the philosophers Theophrastus and Pliny were familiar with their characteristic reddish-purple dye. By the 17th and 18th centuries the so-called 'weed' became an article of international exchange comparable

LEFT The candle lichen, *Lichen candelarius* (*Xanthoria candelaris*), used for colouring animal fat to make candles in Sweden (from J. P. Westrings, *Svenskalafvarnas färghistoria* [1805–9]). Examples of wools dyed with lichens (right).

in scale to oriental spices. One-hundred-and-fifty years ago, botanist William Lindsay, who was particularly interested in the commercial aspects of lichens, which then frequently fetched up to £1000 a ton on the London market, suggested:

> If commanders of ships were aware of the value of these plants, which cover many a rocky coast and barren island, they might with a slight expenditure of time and labour bring home with them such a quantity of these insignificant plants as would realise considerable sums, to the direct advantage of themselves and the shipowners; and consequently to the advantage of the State.

Gathering lichens for the dye trade developed into a massive cottage industry – lichens being collected from remote, often precipitous and dangerous regions in Scotland, Scandinavia and the Canary Islands. People scraped them off the rock with metal hoops, spoons or, if they were poor, seashells. The scale of operations was impressive, even by today's standards. A factory in Glasgow in the 1850s covered an area of about 17 acres (6.9 ha) processing about 254 tonnes (250 tons) of lichen each year. Lichens were steeped in ammonia produced by distilling human urine collected from the suburbs of Glasgow. About 9,100–13,600 litres (2,000–3,000 gallons) were needed every day. The lichen *Rocella tinctoria* was used to manufacture the acid-base indicator litmus paper at the Messrs Johnson of Hendon and Radlett plant in the UK until the 1960s.

BELOW **Cudbear, *Ochrolechia tartarea*, was once an important lichen for dyeing. It contains gyrophoric acid, which decomposes to form orcin. In the presence of ammonia, orcin produces a purple dye (orcein).**

## Perfumes, cosmetics and medicines

Since the 16th century lichens have been used as raw materials in the perfume and cosmetic industries. *Evernia prunastri* and *Pseudevernia furfuracea* are most frequently used, being traded as 'oak moss'. Principally gathered by peasants in southern France, Morocco and the former Yugoslavia, it is estimated that in 1980 alone 8000–9000

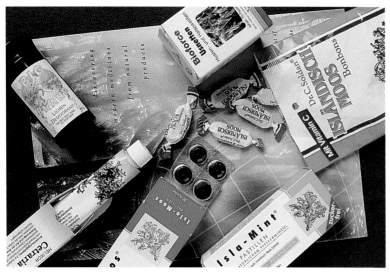

TOP 'Oak moss' *Evernia prunastri*, forms the basis of certain exclusive perfumes (because of its musk-like fragrance and fixative properties).

BOTTOM A selection of lichen medicinal products. Lichens are also often used in potpourri.

tonnes (7870–8860 tons) were collected. Extraction with solvents produces an essential oil or 'concrete', which is highly valued for both its musk-like fragrance and fixative properties for other fragrances. Lichens are often used in potpourri available from major stores worldwide.

According to the 'Doctrine of Signatures', the Creator marked those plants suitable for treating diseases by a resemblance to a specific part of the body. Thus several lichens are illustrated in early herbals, for example *Gerard's Herball* (1597). *Lobaria pulmonaria*, which resembles the lobes of a lung, was used for treating respiratory disorders, and hair moss was thought to be effective against disorders of the scalp. One of the more bizarre beliefs was that lichen growing on human skulls was worth its weight in gold as a cure for epilepsy. There is not much evidence that prescribing these drugs was a success. However, today they still have a firm following as folk medicines, natural remedies and homeopathic aids and are listed in the standard pharmacopoeias present in every chemist shop in the UK.

Several lichen acids, particularly usnic acid, have been found to have antibiotic properties. This is thought to be the reason that lichens were used for babies nappies in New Zealand in the past. A number of lichen-based creams, shampoos, deodorants, cough mixtures and pastilles are marketed today.

Commercial companies in the UK and Canada are currently screening hundreds of lichen-forming fungi for novel pharmaceutical products. The fungal partner

RIGHT 'Iceland moss', *Cetraria islandica*. This lichen is the principal ingredient of many throat pastilles, herbal sweets and herb teas.

is isolated, grown in batch liquid culture and evaluated as a source of commercially exploitable metabolic products. Genetic engineering is also being undertaken by inserting lichen DNA fragments into surrogate bacterial or fungal hosts, for example readily cultured *Aspergillus* spp., to produce novel chemicals.

A few lichen substances are responsible for causing disease including contact dermatitis and eczema, skin complaints that are characterized by reddening and itching. Forestry and horticultural workers are particularly at risk.

# Bioremediation

Slag heaps, rubbish tips, illegal dumps and contaminated land conjure up a nightmarish vision of the 21st century. Virtually all our planet's surface is contaminated to some degree by airborne pollutants from both man-made and natural (for example volcanic) sources. Researchers worldwide are studying microorganisms in strange environments in the search for novel, economical methods of cleaning up the environment. The value of lichens as indicators of metals released to the atmosphere from industry has already been discussed (see p. 80). Indeed, occasionally lichen studies have been made a precondition for granting industrial waste permits for electricity generating stations.

The process of absorption of metals by living or dead biomass (biosorption) is an area of increasing interest for removing potentially toxic and/or valuable metals from contaminated effluents prior to discharge. For instance, an experiment using a new type of

TOP **Contaminated land, Zlatna smelter, Romania.** Only the most tolerant lichens are capable of growing where both high levels of sulphur dioxide and metals are present. Many occur on concrete, which neutralizes the acidity.

BOTTOM **Coniston Copper Mines, Lake District, England.** Site of Scientific Special Interest and of five new British lichens, including copper lecidea (*Lecidea inops*), which is protected under the UK Wildlife and Countryside Act, 1981.

sorbent using dried lichen and seaweed trapped in silica gel recovered almost 100% of metals from deionized water spiked with copper, lead, cadmium and zinc. The performance of these biomass-based sorbents compares favourably with a commercial chelating resin. In addition they are stable and reusable. But because lichens grow slowly, are relatively difficult to culture and, increasingly, it is unacceptable to harvest them on a commercial scale, they are unlikely to be used directly in remediation programmes. But recent advances in biotechnology are encouraging. For instance, economically interesting compounds of slow-growing species can be produced in large quantities by gene transfer into suitable microorganisms.

Many lichens grow directly on a range of toxic substrates, including lead, zinc, arsenic, copper and uranium minerals. They provide model systems for studying the stability of minerals and other toxic compounds in the environment and whether they are likely to be accumulated in the food chain. In some cases, oxalates of toxic metals such as copper, manganese, magnesium, lead and zinc may occur in lichens and on rock surfaces. These insoluble compounds represent one way of locking up toxic metals, enabling lichens to colonize contaminated sites. Considering the chemical diversity of lichens, other mechanisms undoubtedly occur. These will require study using a wide range of analytical techniques at the disposal of scientists today, from light microscopes to massive synchrotron radiation sources, traditionally the tools of physicists.

BELOW LEFT **Natural tolerance mechanism. Moolooite (copper oxalate) crystals coating hyphae within the medulla of *Acarospora rugulosa*. This sample contained 16% dry weight of copper; most copper is fixed outside the cells.**

BELOW RIGHT ***Acarospora rugulosa* growing on the copper mineral, brochantite, Ramundberget, Sweden.**

# Practical projects

## Collecting lichens

Lichens are convenient organisms to study because, with few exceptions, they occur throughout the year. In deciduous woodlands, the best time to study them is when leaves have fallen from trees and light levels are optimal. Most lichens grow very slowly so it is advisable to limit collecting to small samples necessary for identification.

A x10 hand-lens is an essential piece of equipment and can be tied to string and hung around your neck to save it being lost. Other essential items are a suitable shoulder bag providing ready access, a non-folding stout knife and lichen chemicals. Specimens on bark or wood can be removed with a knife, taking care not to damage living tissues. Pruning secateurs are useful for sampling small twigs. Collecting lichens on rocks and other hard surfaces requires the use of a geological hammer and a masonry chisel. Wrap samples in clean tissues and put them in a paper bag, taking care to note down location and habitat details at the time of collection. Moist lichens deteriorate quickly when left in plastic bags for any length of time so their use is to be discouraged.

Please remember to obtain prior permission from landowners and do not collect rare and protected lichens. Avoid damaging buildings, churchyard memorials and other man-made structures. Where necessary, you can make microscopic preparations in the field by sectioning fruiting bodies using a razorblade and mounting directly on to a slide.

Back at home or your hotel it is important to take your samples out of their packets as soon as possible so that they can dry. Dry lichens are easy to maintain and are remarkably resistant to pests. Herbarium samples often look remarkably like living specimens, but over time colours may fade. The major problem is excessive air humidity because many lichens are adapted to absorbing water directly from the air. Under

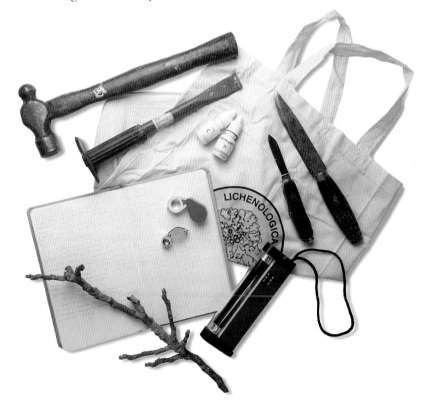

BELOW **Field equipment: stout hammer and chisel (essential for collecting lichens on rocks), knives, collecting bag, hand-lens (x10 and x20), spot-test reagents (K and C), and a portable UV-light, useful for testing fluorescence.**

J.A. Elix: LICHENES AUSTRALASICI EXSICCATI

No. 124 *Teloschistes chrysophthalmus* (L.) Th. Fries
in Gen. Heterolich. cur. recog. 51. [861]
AUSTRALIA. South Australia. Kangaroo Island: King George Beach, 35°40'S,
137°04'E. Growing on dead *Epacris* along the foreshore, elevation 2m.
30 October, 1985                    Leg. J.A. Elix 19757 and L.H. Elix
Chemistry: parir          'major), teloschistin, fallacinal, parietinic acid by HPLC. Anal.
G.A. Jenkins

TOP **Herbarium lichen packet.**

RIGHT **Close-up of** *Teloschistes chrysopthalmus* **shown in the herbarium packet above. Some lichens look virtually the same as the dried specimen in the herbarium, as in the field.**

such circumstances lichens may become attacked by moulds and will deteriorate rapidly unless preventive measures are taken. In lowland tropical regions air-conditioning of the storage place is, therefore, very important. Store samples loose in packets to preserve characters on both upper and lower surfaces, which are important for identification. Specimens on rocks may be glued to pieces of card or wrapped carefully in tissues and placed in small boxes to prevent abrasion. It is important to record full collection details, in case you or someone else wish to revisit the site at some time in the future. Packets may be conveniently stored in shoe boxes.

Identifying lichens may seem a little daunting at first. The best way to learn is to attend a course or field meeting. With experience many can be identified in the field, but it is often necessary to make 'squashes' of lichens and examine them under a microscope to study their spores and other internal features. Chemical spot tests using the standard reagents, K (potassium hydroxide), C (domestic bleach) and PD (*p*-phenylenediamine) are also very useful. Further details are described in all identification texts.

# Assessing the impact of air pollution

High lichen diversity and good growth rate are indicators of cleaner air, compared with a similar habitat with low lichen diversity and damaged lichens. As with any monitoring survey, it is important to consider what you are trying to monitor, other factors that may have an effect and how these can be minimized in the survey, and to select an appropriate method. For results to be statistically significant you need to study a number of samples at more than one site. Improvements in air quality in many industrialized regions have allowed lichens to recolonize and and so the simple 'lichen scales' and 'wall charts' of indicator species are now of limited value. There are three basic ways of monitoring the effects of air quality using lichens: (a) distribution mapping studies, (b) photographic monitoring and, if the equipment is available, (c) chemical analysis.

## Mapping distributions

The following is based on a method originated in Switzerland and subsequently adapted and used successfully in many regions of Italy, largely through the pioneering efforts of Professor Pier Luigi Nimis at the University of Trieste. The full system has been adopted by law by the National Agency for the Protection of the Environment as an integral component of a biological method for monitoring the effects of air pollution throughout Italy. This is an exciting project and one that will involve hundreds of people.

Lichen biodiversity index

Lung cancer mortality

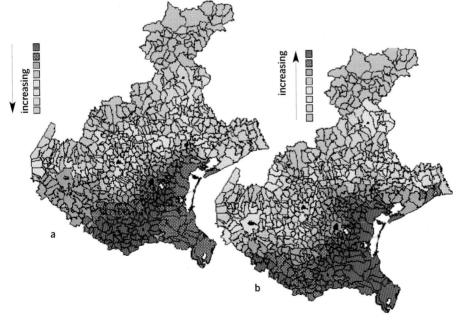

LEFT **Lichen biodiversity (a) calculated by adding up the frequency values of all lichens recorded in a sampling grid of 10 units, and (b) lung cancer mortality in young male residents (expressed as observed/expected cases x100) in the region of Veneto, Italy.**

Carrying out a study
• Choose five mature trees of the same species with straight trunks and circumferences greater than 70 cm (28 in). Avoid leaning trees and excessively mossy trunks.
• Use a sampling grid of 30x50 cm (11.8x19.6 in) subdivided into 10 rectangles of 10x15 cm (3.9x5.9 in) each. Position sampling grid on the trunk where there is highest lichen coverage, at a height of about 1.5 m (5 ft) above ground. Record the aspect (N, NE, NW etc) and be consistent with your recording, selecting similar aspects each time. Remember that different comminities may grow on shaded and sunny sides of trees. Your grid will curve round the tree slightly.
• Identify all lichens in the grid and record the frequency of species within the grid, i.e. give a score of 1–10 for each species, depending on how many of the rectangular subdivisions it occurs in. Seek help where necessary with identification of lichens from a local expert.
• Calculate a lichen biodiversity index by adding together the frequency values of all lichens in the grid on each tree and dividing by 5 (the number of trees examined) to obtain the average for the site.
• Repeat the process at different sites along a transect from a clean to a polluted region or, if time allows, at sites marked on a 5 km$^2$ (2 mile$^2$) grid.
• Plot the lichen biodiversity index on map, and, if possible look for correlations with other pollutants.

A study in the Veneto region of Italy based on 2425 measurements of lichen biodiversity at 662 locations concluded that there was a direct negative relationship between lung cancer mortality and lichen biodiversity. Relatively low levels of sulphur dioxide were recorded in the area, which suggests that sulphur dioxide is not responsible for causing cancer. It is more likely that other substances associated with industrialization may be having an effect.

## Photographic monitoring

Photographic monitoring is great fun and can also be very rewarding scientifically. This is something that anyone who has the necessary patience can do. A series of photographs can be built up over a period of time and the changes interpreted. A photographic record provides detailed information about the type and health of lichens growing in a particular environment and can turn up surprisingly interesting results. We know relatively little about how fast lichens grow and how different species interact with one another under different environmental conditions. Monitoring may be developed further by recording and/or analysing associated environmental factors, for example bark pH, light, rainfall, management.

It is important to be able to relocate accurately the lichen(s) you are monitoring. A convenient method involves using frames (quadrats). Quadrats can be made using wood or a non-rusting, light metal (for example aluminium), ideally in the same 1:1.25 ratio used in 35 mm photography to fill the frame for convenience. Each time you take a photograph, position the camera in a fixed position relative to the lichens to avoid problems of parallax and

ABOVE **Dr Oy Kanjanavanit** showing a park ranger how to monitor lichen communities in Huay Kha Khaeng Wildlife Sanctuary, Thailand.

RIGHT **Monitoring** lichen communities in fire-prone savannah forests in Thailand using a flexible plastic quadrat, with a Kodak colour code to allow colour evaluation of the photograph.

to make it easier to compare results over a period of time.

Various sizes of quadrat may be used according to the size of lichen thalli and the composition of communities being monitored. Clearly, large quadrats are no use on thin branches and small quadrats cannot be used for monitoring large lichens. A colour-test patch, a scale and two lugs are also often fixed to the quadrat to test for colour reproduction, to calculate thallus growth and to facilitate accurate positioning of the quadrat, respectively. Quadrats are not left in the field because this would alter the growing environment of the lichens but may be conveniently relocated (on trees) by means of driving stainless steel screws into the tree aligned with the two lugs at the base of the quadrat. It is important to avoid using galvanized and brass screws, which might produce toxic zinc and copper run-off. A third lug may be usefully located at the top of the quadrat but this is normally only temporarily fixed to the tree using a pin.

Advances in computer and, particularly, digital technology are opening up exciting new possibilities to both capture images conveniently and to manipulate these using imaging processing packages to assess growth changes. It is important to remember that the tree is growing too and that horizontal expansion of the trunk may stretch lichens growing on bark by 1% or more each year. The possibilities for this kind of research are virtually endless. Growth of selected species may be compared between sites and correlations made with pollutants and other environmental parameters.

If you don't have a camera, an even simpler method is to trace the outline of a lichen on transparent plastic sheets and repeat the process after a few months or years, taking care to position the sheet using permanent orientation markers. You can then trace the outlines of the thalli on to paper, cut them out and weigh the pieces of paper (provided you have a sensitive weighing balance) to calculate relative changes in growth. Remember to use the same kind of paper each time.

## Chemical analysis

If you have facilities for chemical analysis, a lichen study can be an excellent way to find out which metals are present in the environment and how far they are dispersed. Environmental pollutants that may affect lichens often contain metals, which can act as signature elements for other pollutants (for example lead for car exhausts by roads). Normally only small amounts of lichen material are required for analysis (around 0.25 g or 0.009 oz) so where macrolichens are plentiful this is not a problem. Remember to sample and analyse a single species because different species accumulate metals to different extents. It is also important to include more than one individual in your sample for analysis to allow for variation between individuals. It is important to select material of a similar age because older parts normally contain higher metal contents than young parts. In leafy *Parmelia* species, sample the outer margin because this represents the current year's growth and is also often easier to separate from its substrate. Remember to clean samples carefully under a binocular microscope. In extremely polluted environments where lichens are scarce, lichen transplants are a convenient practical alternative. You can either suspend branches with shrubby lichens collected from clean environments or use small nylon mesh bags (for example from a nylon stocking) containing lichen fragments and see what they accumulate. Remember, again, to select the same species for your experiment. Your local or national herbaria may have surplus material available for analysis and this can provide baseline information on the previous levels of metals in lichens. Remember to seek permission because some historical specimens are irreplaceable and should not be destroyed.

Particulate contaminants can also be studied on minute fragments of lichen using a scanning electron microscope (SEM), and microprobe.

## Churchyard projects

Churchyards are ideal places to learn about lichens and their ecology. They are also useful habitats for assessing overall air quality if you can compare churchyards in different areas. The thing that makes churchyards so useful is the multitude of conveniently dated surfaces. So it is easy to tell the maximum age of any lichens colonizing their surface. In

BELOW **Lichens may highlight lettering on some gravestones, as does the bright yellow** *Candelariella vitellina* **on this acid gravestone. Pembrokeshire, UK.**

England alone there are around 20,000 churchyards, each taking up roughly a 0.2 ha (0.5 acre) of land. In 1990, The British Lichen Society set up a Churchyards Project, co-ordinated by Tom Chester, to record churchyard lichens throughout the UK. An annual weekend course is given to instruct beginners in the identification and ecology of churchyard lichens. So far, over 60 lowland churchyards and the precincts of two lowland cathedrals (Salisbury and Winchester) have each been found to contain more than 100 species. Similar projects are now starting in Germany and the Netherlands.

Churchyards provide an incredible range of habitats for study. North-, south-, east- and west-facing vertical gravestones provide different aspects for colonization. Horizontal chest tombs are also usually present. Several different rock types may be found, including granite (rough and polished), marble, limestone, sandstones and slates. Lichens may also grow on the often very ancient boundary fence and on trees. The possibilities for simple, yet highly informative studies are virtually endless.

# Ideas for lichen projects

## ● How fast do lichens grow?

Compare the same lichen species on gravestones of different ages and work out the growth rate. You can do this by measuring the greatest diameter of thalli of a particular species on as many dated stones as possible and plotting a graph of size against time. Remember to select several examples of the lichen on each stone and calculate an average. You can extend the study by either tracing or photographing thalli and repeating this exercise the following year. You could then compare actual changes in one year with the changes predicted by your graph.

## ● What do lichens grow on?

Compare the same species on different gravestones and assess whether the type of substrate (granite, marble etc) is important. Do particular lichen species prefer particular aspects? Is the age of the gravestone important?

## ● How do lichens change over time?

Compare the lichen communities in a churchyard on a single type of gravestone (marble, sandstone or granite etc). Do similar lichen communities occur on stones of different ages? What other factors may be responsible. Record how the lichens change over time either using a quadrat grid or by photography.

## ◐ What happens when two lichens meet?

Study several examples where the same two lichen species meet to assess what normally happens. There are several possibilities: (a) one lichen can overgrow and destroy the other; (b) they can both stop growing; (c) they can join together to make what appears to be a single lichen thallus. Try to work out what is happening from your observations. Better still repeat your survey next year and see whether your original hypotheses were correct. Does it matter if the lichens are young or old? Are particular species consistently more aggressive than others, and if so does this depend on their growth rate or on some other factor? Do leafy and crustose lichens behave differently?

## ◐ Assessing pollution levels

If you examine pollution levels by comparing lichens in churchyards in different areas, i.e. along a transect from a clean to polluted zone, can you detect any changes in the types and abundance of lichens present? Choose a single easily recognizable species and try to discover if its habitat remains constant in all churchyards or if there are differences between sites. Is the variation significant and does this relate to air quality or some other factor? Are there any differences in the colonization patterns on gravestones of different ages? If so, can you explain why? Obtain pollution data for your area and look for correlations with your lichen data.

## ◐ Do lichen species that are tolerant to certain metals occur?

Metals are present in all churchyards. Lead engravings are used on gravestones for lettering and copper plaques are often inserted on tombstones. Stained glass windows contain lead between the panes and the windows are often protected by grilles containing copper or iron. Copper lightning conductors are also often present. You can tell where copper is present because blue-green secondary copper compounds are fixed, especially where mortar is present. Are metal-tolerant lichens present restricted to a particular type of metal-rich run-off, for example lead, copper or iron? If you have the facilities you can analyse lichens chemically, identify where metals are localized and suggest how they are fixed. Scanning electron microscopy and light microscopy are basic techniques that can open up fascinating research opportunities.

# Glossary

**Actinorrhiza** Association of bacteria with the roots of higher plants.

**Allelopath** A chemical released by a lichen (or a plant), which inhibits the growth of other lichens or plants and reduces competition.

**Apothecium (pl. apothecia)** A more-or-less flat, cup- or saucer-like fruiting body.

**Ascomycete** Class of fungi to which the morel and orange peel fungus belong.

**Ascospores** Sexual spores formed by ascomycetes.

**Ascus** The sac-like cell that contains the ascospores.

**Basidiomycete** Class of fungi to which the familiar mushrooms and toadstools belong.

**Basidium (pl. basidia)** The characteristic cell of the basidiomycete fungi, which produces four reproductive spores at its tip.

**Becquerel (Bq)** The amount of a radioactive substance normally expressed in terms of its radioactivity rather than its mass. One Becquerel = 1 disintegration per second.

**Cephalodium (pl. cephalodia)** A wart-like structure produced in or on a lichen containing cyanobacteria as well as green alga.

**Chemosystematics** The use of chemical characters for classification.

**Conidium (pl. conidia)** An asexual spore.

**Cortex** A protective layer of compacted fungal cells usually present on the upper surface above the photobiont layer and occasionally also on the lower surface.

**Cyanobacterium (pl. cyanobacteria)** Formerly called blue-green alga(e).

**Cyphellum (pl. cyphellae)** A round, cup-like break in the lower cortex.

**Fruiting bodies** Round or stretched structures producing fungal spores, which may form a new lichen.

**Haustorium (pl. haustoria)** Hyphae that clasp and invade algal cells and are thought to play a role in transferring carbohydrate to the fungus.

**Heterotrophy** In which an organism depends for its nourishment on organic matter already produced by other organisms.

**Hypha (pl. hyphae)** A fungal filament.

**Isidium (pl. isidia)** A peg-like outgrowth of a lichen containing fungus and alga, which if broken and transported to a suitable location will grow into a new lichen.

**Lichen substances** Chemicals produced by lichens; many are unique.

**Lichenization** The process by which a fungus meets a compatible alga to form a lichen.

**Macrolichen** A large, leaf-, shrub- or beard-like lichen.

**Mazaedium (pl. mazaedia)** Powdery, black sooty-like spore mass.

**Mechanical hybrid** Where hyphae from one lichen grow into another lichen so that tissues of both lichens are present in a single individual.

**Medulla** The central, air-filled region of a lichen composed of loose fungal hyphae.

**Microcrystallization tests** Tests to identify lichen substances based on crystal formation.

**Mycelium (pl. mycelia)** A mass of hyphae.

**Mycobiont** The fungal partner in a lichen.

**Mycorrhiza** Association of a fungus with a higher plant.

**Nitrophilous** Lichens that prefer habitats rich in nitrogen.

**Perithecium (pl. perithecia)** Flask-like fruiting body in which the spore-bearing layer (hymenium) is enclosed apart from a small opening at the tip.

**Photobiont** A green algal or cyanobacterial partner occurring in lichens.

**Photomorphs** The same lichen containing either a cyanobacterium or a green alga.

**Photosynthesis** The process by which green plants containing chlorophyll use the energy of sunlight to produce carbohydrate.

**Polyphyletic** Derived from two or more ancestral lines. Lichen fungi do not share a common ancestor and have evolved on at least four separate occasions.

**Pseudocyphellum (pl. pseudocyphellae)** An area where the medulla comes to the surface. It may be round, stretched or irregular in shape. It is thought to be important to allow gases to enter the lichen thallus.

**Rad** A unit for measuring absorbed energy from radiation (gamma rays, X-rays etc).

**Respiration** The production of energy through breakdown of carbohydrates.

**Rhizines** Root-like hairs or hyphae, which attach the lichen to its substrate.

**Soralium (pl. soralia)** A structure or region of a thallus bearing soredia.

**Soredium (pl. soredia)** Powdery mixtures of fungal and algal cells, which serve in reproduction.

**Species pairs** Identical species, one of which is fertile (the primary species) and the other reproduces vegetatively (the secondary species).

**Spores** Sexual reproductive cells produced in an ascus or as a basidium.

**Symbiont** A partner living in symbiosis.

**Symbiosis** The living together of unlike organisms; it may be for mutual benefit or not.

**Thallus (pl. thalli)** The lichen body containing the fungus and photobiont layers.

**Thin layer chromatography (TLC)** Technique used to identify lichen substances using an aluminium or glass plate coated in silica on which acetone extracts of lichens are spotted. The substances in the spots are separated in organic solvents.

**Trichogyne** The female hyphal cell that accepts male cells (conidia) in sexual reproduction.

# Index

Italicised numbers refer to captions.

*Acarospora* spp. 54, *54, 55, 73, 81, 98*
Acharius, Erik 48
acid rain 80, 84
*Alectoria* spp. *58*
*Aleuria aurantia 8*
algae 5-8, 9-15, 21, *22, 23,* 48;
    in extreme temperatures 69,
    73; in filamentous lichens 45;
    in fossil lichen 46; in jelly
    lichens 17; in mushroom
    lichens 35; and nutrition 26-8,
    49; in rain forest species 60;
    in script lichens 43
alpine habitats 67-9
ancient monuments 53
Antarctic 69-73, *72, 75*
Arctic *22,* 49, 66-7, 82-3
*Arctoparmelia centrifuga 38, 38*
*Arthonia tavaresii 45, 45*
*Aspergillus* spp. 97
*Aspicilia* spp. 28, 53

biodiversity 33-45, 56, 60, 84;
    indices 77-8, *101,* 102
biomonitoring 76-87, *76*
bioremediation 97-8
birds *22,* 50, *50,* 59; droppings
    27, 28, 64, 69, 70
British soldiers lichen *39*
*Bryoria* spp. 50, *58,* 66, 70, 80
*Buellia* spp. 73
*Byssoloma discordans* 61

*Calicium* spp. 36, *36, 37*
*Caloplaca* spp. 53, 65, *65;*
    *C. flavescens* 44, *45; C.* cf
    *ignea 54, 55; C. luteoalba* 56
camouflage 51
*Candelariella* spp. 28
candle lichen *93*
carbon fixation 8, *25,* 26
cephalodia 10, 27

*Cetraria islandica* 96
*Chaenotheca* spp. 36, *37*
chemical analysis 31, 48, 88-
    90, 104-5
chemical compounds *see*
    substances
Christmas lichen 44, *44*
churchyards 33, 44, 62-4, *62,*
    *63;* use in dating studies 45,
    91, 105-6, *105*
*Cladina* spp. 92
*Cladonia* spp. 5, 10, 39-42, 49,
    92; *C. chlorophaea 41;*
    *C. cristatella 39; C. deformis*
    *39; C. floerkeana 41;*
    *C. rangiferina 40, 66;*
    *C. stellaris 39*
classification 8, 13, 18, 33, 37,
    47-8
clean air act 79
coastal habitats 64-5, 73-5
*Coenogonium* spp. 6, 45, *45*
collecting 99-100, *99, 100*
*Collema* spp. 9, *17*
commercial applications 92-8
conservation 62, 83-4, 86-7,
    92, 101
copper lecidea 54, 97
coral lichen 36, *36*
cryptoendoliths *72,* 73
*Cryptothecia rubrocincta* 44, *44*
cudbear 94
culturing lichens 8-9, 10, 28,
    92, 98; *Aspergillus* spp. 97;
    *Xanthoria* spp. 9
cyanobacteria 5-6, 10-11, 15,
    49, 60

Darwin, Charles 71
dating surfaces 90-1
defence mechanisms 31, 50, 71-2
*Dendriscocaulon* spp. 13
desert habitats 24-5, 49, 73-5, *75*
disease 97
*Dictyonema glabratum 35, 35*
*Diploschistes muscorum* 10

*Dirina massiliensis* 24
DNA 15, 24, 36, 46-7, 97
doctrine of signatures 95
dyes 93-94

ecology: role of lichens in 49-55;
    *see also* biomonitoring;
    zonation
economic uses 92-98
*Erioderma* spp. 86-7, *86, 87*
*Evernia prunastri* 81, 94-5, *95*
evolution 46-7

filamentous lichens 45
flask lichens 43
floiicolous lichens 60-61
food: for animals 50; for man 92
forest habitats 8, 33, 36, 43,
    49-50, 56-9; biomonitoring
    of 83-7; *see also* rain forests
fossils 46, *46,* 60, 73

global warming 49
Gondwanaland 60
gravestones *see* churchyards
growth 18, 28, 90-1, 106

habitats 51; dry 24-6, *25,* 28;
    wet 7, 8, 31, 36, 43; *see also*
    alpine, Antarctic, Arctic,
    coastal, desert, forest, island,
    mountain, *and* urban habitats
*Heppia adglutinata* 6
Herbaria 78, 90
humidity 13, 24, 56, 73-5, 86,
    99-100
hybrids 23
*Hypogymnia* spp. *5*
*Hypotrachyna britannica* 37

Iceland moss 96
identification techniques 10, 22,
    31, 32, 36, 41-2
'indicator' species 56
island habitats (Atlantic) 13, 22,
    37, 45, 65, 84

Iwatake 92

jelly lichens 9, 17

*Lecanactis* spp. 56, 62
*Lecanora* spp: *L. cascadensis* 88;
    *L. conizaeoides* 78, *78,* 80;
    *L. esculenta* 75; *L. muralis*
    28, 44, *45; L. polytropa* 67;
    *L. vinetorum* 64
Lecanorales (order) 47
*Lecidea* spp. 73; *L. inops* 54,
    97; *L. lactea* 52, 88, 89, 90;
    *L. lapicida* 88, 90; *L. theiodes*
    88-90
*Lepraria* spp. 22
*Leptogium* spp. 9, 56
*Letharia vulpina* 31
*Lichen candelarius* 93
life cycle *23*
light sensitivity 7, 13, 18, 24,
    28; desert habitats 75; forest
    habitats 56, 60; UV light 31,
    39-41, 70-2
*Lobaria* spp.13, 78, 80; *L.*
    *pulmonaria* 56, 83-4, *83*
longevity 19, 26
long-beard lichen 56
*Loxospora elatina* 21

map lichen 53, *53*
Mars 73
medicines 92-3, 95
metals 64, 88-90, *89;*
    contamination by 50, 62,
    80-1, *81,* 97; monitoring
    levels of 37, 77, 104-5;
    toleration by lichens 31, 52,
    53-4, 97-8, 107
Methuselah's beard 56
minerals *see* metals; rocks
mountain habitats 22, 35, 65,
    67-9, *67, 72*
*Multiclavula vernalis* 35, *35*
mushroom lichens 34-5

*Nephroma* spp. 13, 60
*Neuropogon* see *Usnea aurantiaco-atra*
nitrogen 75, 80, 83; fixation 7, 28, 49-50, 60, 69, 78
*Nostoc* (algae) 10, *11*, 17, 60
nutrition 26-8, 61; *see also* nitrogen fixation, photosynthesis

oak moss *81*, 94-5, *95*
*Ochrolechia tartarea* 94
oil slicks 65
*Omphalina umbellifera* 34, *34*
*Opegrapha filicina* 42
orchella weed (*Roccella* spp.) 75, 93, 94
ozone levels 70-1

*Parmelia* spp. 37-8, *37*, 53, 104; *P. caperata* 16, 90; *P. sulcata* 31
*Parmentaria chinensis* 43
particles 80-81
*Peltigera horizontalis* 56
perfumes 92, 94, 95
*Phaeographis* spp. *42*
photobionts *see* algae and cyanobacteria
photographic monitoring 102-4, *103*, 106
photomorphs 10-15, *13*
photosynthesis 5, 7, 25, 26, *26*; disruption of 31, 78; effects of temperature on 26, 72, 73, 75
*Physcia* spp. 23, *23*, 31, 32; *P. adscendens* 56; *P. aipolia* 28, *29*
physiology 24-8
pin lichens 36, *36*, 37
pixie-cups 39, 41
*Placopsis gelida* 27
pollution 50-1, 76-87, 97-8, 101-5, 107; toleration by lichens 22, 33, 26, 36, 52; effects on propagation 44, 45
pollution lichen 78, *78*, 80

*Porpidia* spp. 53
predators 50, 66-7
projects: assessing air quality 101-5; churchyard 105-6; collecting specimens 99-100; ideas for 106-7
prospecting 88
*Pseudevernia furfuracea* 80, 94-5
*Pseudocyphellaria* spp. 18, 59, 60, 78
*Psilolechia* spp. *33*, 44, *45*
*Psoroma durietzii* 59, 60
*Pyrenocollema halodytes* 65
pyrenolichens 43
*Pyrenula* spp. 43, *45*

radiation: as investigative tool 98; nuclear 66-7; UV 31, 39-41, 70-2
rain forests 13, *34*, 35, 36, 45, 59-61, *61*
*Ramalina* spp. 65, 75; *R. farinacea* 80; *R. fraxinea* 56; *R. maciformis* 25, 73; *R. menziesii* 90
*Ramonia azorica* 45, *45*
recolonisation 79
reindeer 50, 66-7, *66*, 83
reindeer moss 39-41, 49, 82-3; *Cladina* spp. 92; see also *Cladonia* spp.
remote sensing 82-3
reproduction: asexual 21-2, 38, 39, 69, 79; hybridisation 23-4; sexual 7, 10, 19-21, 61
*Rhizocarpon geographicum* 53, *53*, 90, 91, *91*
*Roccella* spp. 75, 93, 94
rock tripe 67
rocks 51-4, *52*, *53*, *54*, 55, 64-5; calcareous 28, 43, 52, 53, 56; silicate *52*, *53*, 69, 73, *91*; *see also* fossils

script lichens 42-3, *42*
smogs 77
snow beds 69

*Solorina crocea* 69, *69*
species pairs 24
*Sphaerophorus* spp. 36, *36*
sphagnum mosses 35
spores 19-21, 23, *23*, 43, 46; use in classification 36, 48; use in culturing 8-9
spot test reactions 31-2
*Staurothele* spp. 21
*Stereocaulon* spp. 5, 49
*Sticta* spp. 18, 60, 78; *S. canariensis* 12, 13, 14, *15*; *S. felix* 13, *13*
structure 6, 15-18, 26; diagrams/photographs of: apothecia *19*; asci *19*, *37*; ascospores *19*, *23*; cortex *16*, *17*, *81*; cyphellae *18*; hyphae *47*; isidia *22*; mazaedia *37*; medulla *16*, *17*, *81*; perithecia *20*, *42*; pseudocyphellae *18*; soralia *21*, *37*, *38*; soredia *21*, *23*; thalli 7
substances 28-32, 33, 50, 71-2, 82, 90; calcium oxalate 26, 28, *28*, *29*, 52; skin irritation 97
sunscreens 71
symbiosis 5-8

tar-spot lichen 65, *65*
taxonomy *see* classification
*Teloschistes* spp. 74, 75, *100*
temperature 24, 31; effects on photosynthesis 26, 72, 73, 75
*Thamnolia vermicularis* 22, 69, *69*
*Thelomma* spp. *5*
transplanting 84, 105
*Trebouxia* (algae) 9, *10*, 45, 73
tree lungwort *see Lobaria pulmonaria*
trees: ash 80, 84; aspen *29*; box 61; *Elaeocarpus* 61; hazel *43*; holly *43*; juniper 45; laurel 60-1; *Nothofagus* *34*; oak 42, 83; photographic monitoring

of 102-4, *103*; redwood 58; spruce 58, 86
*Trentepohlia* (algae) 9, *11*, 43, 60

*Umbilicaria* spp. 67, 92
unattached lichens 75
urban habitats 33, 36, 38, 44, 62-4, 77-81
*Usnea* spp. 5, 17, 70, 79, *79*; *U. articulata* 50, 78; *U. aurantiaco-atra* 70, 71-2, *71*; *U. inflata* 17, 56; *U. longissima* 56, 58-9, *58*, 85
uv radiation 31, 39, 71-2

*Verrucaria* spp. 43, 64, 65, *65*

'Wanderflechten' (unattached lichens) 75
water: absorption of 26, 49, 73; and hydrophobic mechanisms 18, 31; role in physiology 25-6, *25*, *26*; role in reproduction 21; *see also* humidity
weathering (of rocks) 51-3, *52*, *53*, *54*, 73
*Winfrenatia reticulata* 46, *46*, 60
wolf lichen 31
worm lichen 22, 69, *69*

*Xanthoparmelia mougeotii* 38
*Xanthoria* spp. 5, 31, 32, 53, 65, *65*, 79, 80, *80*; *X. candelaris* 93; *X. elegans* 67, 68; *X. parietina* 9, 18, *19*, *20*, 31, 56 (life cycle 23, *23*; and bird droppings *27*, 28)

zonation 64-5, *65*, 73

# Further information

## Selected books

*Die Flechten Baden-Württembergs,*
V. Wirth. Eugen Ulmer, Stuttgart, 1995.
[555 superb colour photos, maps and keys
for the identification of ca 1000 species]

*Lichen Biology,* T.H. Nash (ed.).
Cambridge University Press, Cambridge,
1996.
[useful text on lichen biology for the more
advanced student]

*The Lichen Flora of Great Britain and
Ireland,* O.W. Purvis, B.J. Coppins, D.L.
Hawksworth, P.W. James & D.M. Moore.
The Natural History Museum & British
Lichen Society, London, 1992.
[detailed descriptions of crustose and
macrolichens essential for advanced
lichenologists world-wide]

*Lichens. An Illustrated Guide to the British
and Irish Species,* F.S. Dobson.
Richmond Publishing, 2000.
[popular identification guide to British lichens
covering ca 450 spp., over 2000 colour
photographs, maps and line illustrations]

*Lichens,* O.L. Gilbert. The New Naturalist,
Harper Collins, London, 2000.
[popular authoritative, well-illustrated
account of the ecology of British lichens]

*Lichens of North America,* I.M. Brodo,
S.D. Sharnoff & S. Sharnoff.
Yale University Press, in press.
[popular identification guide with 922
superb colour photographs illustrating
800 species from US and Canada, with
keys or notes covering an additional
700 species; extensive introductory
chapters covering lichen biology,
environmental monitoring etc]

*Lichens of rainforest in Tasmania and
south-eastern Australia,* G. Kantvilas &
S.J. Jarman. Flora of Australia
Supplementary Series 9. The Australian
Biological Resources Study, Canberra.
Australia, 1999.
[over 200 superb colour photographs]

*Macrolichens of the Pacific Northwest,*
B. McCune & L. Geiser. Oregon State
University Press/ U.S.D.A. Forest Service,
Corvallis, 1997.
[Keys for ca 500 species, descriptions and
illustrations (many colour photos) for 210
species; information on lichens and air
quality etc]

*Pollution Monitoring with Lichens,*
D.H.S. Richardson. Naturalists'
Handbooks, 19, Richmond Publishing,
Slough, 1992.
[basic introductory text]

## Some useful web links

British Lichen Society
http://www.argonet.co.uk/users/jmgray/
[information on the British Lichen Society,
list of British lichens and synonyms etc]

Environmental Surveillance from Satellites
http://www.itek.norut.no/itek/sat/projects/
    sur_sat.fm.html
[surveillance of lichen heaths impacted by
air pollution by satellite]

International Association for Lichenology
http://www.botany.hawaii.edu/lichen/
[provides contact details for lichenologists
world-wide]

Lichen Database of Italy
http://biobase.kfunigraz.ac.at/flechte/owa/
    askitalflo
[pioneering database for the enthusiast
enabling search for ecological paramaters etc]

Lichen Information system
http://www.sbg.ac.at/pfl/projects/lichen/
    index.htm
[links to interesting lichen sites]

List server for all lichenologists:
listproc@hawaii.edu.
To sign up:
subscribe lichens-1 then give First name,
Last name.
To mail a message to the lichen list members:
lichens-l@hawaii.edu

The Natural History Museum
http://www.nhm.ac.uk/
[information on lichenology, science and
education]

Search recent Literature on Lichens
http://www.toyen.uio.no/botanisk/
    bot-mus/lav/sok_rll.htm
[useful database for searching for key
words, authors, journals etc]

USDA Forest PNW Lichens and Air Quality
http://www.fs.fed.us/r6/aq/lichen/
[National Forest on-line searchable GIS
database for Pacific Northwest]

NB. Web addresses are subject to change.

# Acknowledgements

## Picture credits

Unless otherwise stated all images copyright The Natural History Museum

Front and back covers, title page, Ian Munroe; Front cover (inset) US edn. only, Laurie Campbell; **p.5** (left and right), Sylvia & Stephen Sharnoff; **p.8** Peter W. James; **p.9** (left), Ian Munroe; **p.9** (right), Rosmarie Honegger; **p.27** (top), Sylvia & Stephen Sharnoff; **p.27** (bottom), Peter W. James; **p.31** Rosmarie Honegger; **p.32** (bottom left), Peter W. James; **p.34, p.35** (top), Bruce Fuhrer, permission from Forestry Tasmania and the Tasmanium Herbarium; **p.35** (bottom), Burkhard Büdel; **p.36** (left & right), **p.37** (top left), Mats Wedin; **p.38** (top), Peter W. James; **p.38** (bottom), **p.39** (top & bottom), Sylvia & Stephen Sharnoff; **p.40** Laurie Campbell; **p.41** (top), Peter W. James; **p.41** (bottom), Laurie Campbell; **p.42** (left, top right), Bryan Edwards; **p.42** (middle right), Robert Lücking; **p.44** Sylvia & Stephen Sharnoff; **p.45** (top right & bottom left), Peter W. James; **p.45** (bottom right), Robert Lücking; **p.46, p.47** (left & right), Tom Taylor **p.49** Peter D. Crittenden; **p.50** (right), **p.51**, Peter W. James; **p.52** (top), William Purvis; **p.53** Martin Lee; **p.54** (top), William Purvis; **p.54** (bottom), **p.55**, Sylvia & Stephen Sharnoff; **p.57** (1st & 4th top, 2nd & 3rd bottom), Peter W. James; **p.58** (right), Sylvia & Stephen Sharnoff; **p.59** (left & right), Bruce Fuhrer, permission from Forestry Tasmania and the Tasmanium Herbarium; **p.61** (left & top right), Angela Newton; **p.61** (bottom right), Robert Lücking; **p.62** (right), **p.63** Barbara Hilton, **p.65**, Peter W. James; **p.66** (bottom), Roger Rosentreter; **p.67** Sylvia & Stephen Sharnoff; **p.68, p.69** (top and bottom), Heribert Schöller; **p.70** Burkhard Schroeter; **p.72** Peter Convey / British Antarctic Survey; **p.74** Peter D. Crittenden; **p.82** (left and right), Hans Tømmervik; **p.83** (left and right), Laurie Campbell; **p.86** Jorn Arne Saether; **p.87** (left), Burkhard Büdel; **p.87** (right), David Yetman; **p.88**, William Purvis **p.91**, Sylvia & Stephen Sharnoff; **p.94, p.95** (top), **p.96**, Ian Munroe; **p.97** (bottom), William Purvis; **p.101** Reprinted with permission from *Nature* 387, 463. ©1997 Macmillan Magazines Ltd and P.L. Nimis.

LINE ILLUSTRATIONS
**p.7** adapted from fig. 3 in V. Ahmadjian, *The Lichen Symbiosis*. John Wiley & Sons, New York, 1993.
**p.13** adapted from fig. 2 in P.W. James & A. Henssen. In *Lichenology: Progress and Problems*, D.H. Brown, D.L. Hawksworth & R.H. Bailey. Academic Press, London, 1976.
**p.18, p.19** (bottom), **p.20** (bottom), **p.21** (right), **p.22** (left), **p.42** (bottom right) adapted from drawings by A. Orange in Purvis, O.W. *et al.* (1992) *Lichen Flora of Great Britain and Ireland*. The Natural History Museum & British Lichen Society, London.
**p.23** adapted from S. Ott, *Bibliotheca Lichenologica* 25: 81-93.
**p.25** adapted from O.L. Lange, *Oecologia* 45: 82-87.
**p.26** adapted from R.Honegger in *The Mycota* V. Part A., Springer, Berlin, 1977.
**p.66** (top) adapted from E. Gaare & T. Skogland. In *Fennoscandian Tundra Ecosystems. 2: Animals and Systems Analysis*, F.E. Wielgolaski. Springer-Verlag, New York, 1975.
**p.77** adapted from D.L. Hawksworth & Rose, *Lichens as Pollution Monitors*. Arnold, London, 1976.

## Author's acknowledgements

I would like to thank Mats Wedin (University of Umeå, Sweden) for drafting sections on 'coral lichens and pin lichens' and 'evolution, classification and naming', Linda Geiser (USDA-Forest Service, Siuslaw National Forest) for information on the US Air Quality Biomonitoring Program and Hans Tømmervik (The Polar Environmental Centre, Tromsø, Norway) for information on satellite monitoring using lichens. Oliver Gilbert (University of Sheffield), Clifford Smith (University of Hawaii) and Peter James (The Natural History Museum) are thanked for critical comments on the manuscript. I am grateful to Tom Chester for assistance with ideas for churchyard projects, to Peter York (The Natural History Museum) for photomicroscopy, Pat Wolseley and Chris Stanley, NHM photographers for expert photographic assistance, and all other lichenologists mentioned above who have so generously provided photos and other images.